AF625165

Sawan S. Sinha

Fundamentals of Fluid Mechanics

Ane Books Pvt. Ltd.

Springer

Sawan S. Sinha
Department of Applied Mechanics
Indian Institute of Technology Delhi
New Delhi, Delhi, India

ISBN 978-3-032-04209-5 ISBN 978-3-032-04210-1 (eBook)
https://doi.org/10.1007/978-3-032-04210-1

Jointly published with Ane Books Pvt. Ltd.
In addition to this printed edition, there is a local printed edition of this work available via Ane Books in South Asia (India, Pakistan, Sri Lanka, Bangladesh, Nepal and Bhutan) and Africa (all countries in the African subcontinent).
ISBN of the Co-Publisher's edition: 978-81-19662-76-0

© The Author(s), under exclusive license to Springer Nature Switzerland AG 2026

This work is subject to copyright. All rights are solely and exclusively licensed by the Publisher, whether the whole or part of the material is concerned, specifically the rights of reprinting, reuse of illustrations, recitation, broadcasting, reproduction on microfilms or in any other physical way, and transmission or information storage and retrieval, electronic adaptation, computer software, or by similar or dissimilar methodology now known or hereafter developed.
The use of general descriptive names, registered names, trademarks, service marks, etc. in this publication does not imply, even in the absence of a specific statement, that such names are exempt from the relevant protective laws and regulations and therefore free for general use.
The publishers, the authors, and the editors are safe to assume that the advice and information in this book are believed to be true and accurate at the date of publication. Neither the publishers nor the authors or the editors give a warranty, express or implied, with respect to the material contained herein or for any errors or omissions that may have been made. The publishers remain neutral with regard to jurisdictional claims in published maps and institutional affiliations.

This Springer imprint is published by the registered company Springer Nature Switzerland AG
The registered company address is: Gewerbestrasse 11, 6330 Cham, Switzerland

If disposing of this product, please recycle the paper.

In the loving memory of my father Shri Amarnath Suman

Competing Interests The author has no competing interests to declare that are relevant to the content of this manuscript.

Contents

1 Tensors ... 1
1.1 Expressing a First-Order Tensor Using a Cartesian Coordinate System ... 2
1.2 Transformation Rule: First-Order Tensors ... 2
1.3 Expressing a Second-Order Tensor Using a Cartesian Coordinate System ... 4
1.4 Transformation Rule: Second-Order Tensors ... 5
1.5 Expressing Higher-Order Tensors Using a Cartesian Coordinate System ... 6
1.6 Einstein's Summation Rule ... 6
1.7 Tensor Operations ... 9
1.7.1 Dot Product of Two Tensors ... 9
1.7.2 Double Dot Product of Two Tensors ... 11
1.7.3 Cross Product of Two Vectors ... 12
1.8 The ϵ-δ Identity ... 13
1.9 Spatial Derivatives of Tensors ... 13
1.10 Index Notation and Tensor Identities ... 15
1.11 Expressing Tensors Using the Cylindrical-Polar Coordinate System ... 19

2 Mechanics: A Brief Review of the Fundamentals ... 23

3 Description of Fluid Motion ... 27
3.1 The Continuum Description ... 27
3.2 The Lagrangian Description of Fluid Continuum ... 29
3.3 The Eulerian Description of Fluid Continuum ... 30

4 Fluid Kinematics ... 35
4.1 Streamlines, Streaklines, and Pathlines ... 36
4.2 A Fluid Element ... 38
4.3 A Deforming Fluid Element in a 2C, 2D Velocity Field ... 40
4.4 A Deforming Fluid Element in a 3C, 3D Velocity Field ... 46

5 Governing Equations of Fluid Motion ... 51
5.1 Flux Through a Surface ... 51
5.2 A Control Volume and the Reynolds Transport Theorem ... 53
5.3 The Differential Form of RTT ... 55
5.4 The Governing Equation of ρ ... 59
5.5 The Governing Equation of $\underline{V}$ ... 60
5.5.1 The Stress Tensor ... 61
5.5.2 External Force on a Fluid Element ... 66
5.6 The Governing Equation of T ... 71
5.7 Closure of Governing Equations ... 78
5.7.1 Symmetry of the Stress Tensor ... 79
5.7.2 A Closure Equation for the Stress Tensor ... 80
5.7.3 A Closure Equation for the Heat Flux Vector ... 82
5.8 The Navier–Stokes Equation Set ... 82
5.8.1 Incompressible Navier–Stokes Equation Set ... 84

6 Interaction of a Flow Field with a Solid Surface ... 87
6.1 Velocity Constraints on a Flow Field on a Solid Surface ... 87
6.2 Force System Exerted by Fluid on a Solid Surface ... 88

7 Hydrostatics ... 93
7.1 Net Hydrostatic Force on a Wetted Solid Surface ... 94
7.2 Resultant Hydrostatic Force System on a Submerged Solid Body ... 97

8 Some Simple Representative Flows of an Ideal Fluid ... 103
8.1 Governing Equations of an Ideal Fluid ... 103
8.2 Bernoulli's Equation ... 104
8.3 The Vorticity Equation ... 109
8.4 Potential Flows ... 111
8.5 Potential Flows with 2C, 2D Velocity Fields ... 114
8.6 A Doublet ... 117
8.7 Solution of Potential Flow Past a Long Circular Cylinder ... 120
8.8 Solution of Potential Flow Past Bodies of Arbitrary Shapes ... 126

9 Some Simple Representative Flows of a Viscous Fluid ... 127
9.1 Flat Plate Boundary Layer ... 127
9.1.1 Order-of-Magnitude Analysis ... 130
9.1.2 The Continuity Equation ... 132
9.1.3 The V_1 Equation ... 133
9.1.4 The V_2 Equation ... 136
9.1.5 Stream Function ... 140
9.1.6 Blassius Solution ... 144
9.2 Boundary Layer Separation ... 155
9.3 Internal Flows and Fully Developed Velocity Profiles ... 159

10 Governing Equations of Fluid Motion in Non-inertial Reference Frames 165
10.1 A Non-inertial Reference Frame 165
10.2 A Frame Translating with Constant Velocity 170

References 173

Index 175

About the Author

Dr. Sawan S. Sinha is a professor in the Department of Applied Mechanics, Indian Institute of Technology (IIT) Delhi, India. He received his B.Tech. degree in Aerospace Engineering from IIT Kanpur (2001) and his Ph.D. degree from Texas A&M University (Aerospace Engineering), College Station, TX, USA (2009). Besides being a faculty member at IIT Delhi, his professional experience from the past includes working as an engineer for Defence Research Development Organization (DRDO) during 2003–2004 and later as a post-doctoral Research Associate at Texas A&M University (2010–2012).

Dr. Sinha has more than a decade-long teaching experience at IIT Delhi. He frequently teaches courses on Fluid Mechanics, Engineering Mechanics, Turbulence and its Modelling and Computational Fluid Dynamics, as well. Based on students' feedback, he has received Awards for Teaching Excellence at IIT Delhi (Turbulent Flows, 2013 and Fluid Mechanics, 2014).

Chapter 1
Tensors

A *tensor of order n* ($\in \{0, 1, 2, 3, ...\}$) in a three-dimensional space is a mathematical object comprised of 3^n numbers, the numerical value of which may depend on the coordinate system we choose to express that tensor. These 3^n numbers are called the *scalar components of the tensor in the chosen coordinate system*. Even though these scalar components, in general, are dependent on the choice of the coordinate system used to express it, the tensor as a mathematical entity and the physical quantity that it represents must be independent of the choice of the working coordinate system. In other words, the tensor itself is *invariant* to the choice of the coordinate system being used to express it. This property of the invariance of the tensor is ensured by a set of relationships that exists between the set of 3^n components of the tensor in one coordinate system and another set of the 3^n components of the same tensor in the other coordinate system. Such relationships are called the *transformation rules* of tensors. A tensor of order zero is a special case wherein the transformation rule is trivial. Such a tensor is described by one (3^0) number, which is independent of the choice of the working coordinate system.

In our study of fluid mechanics in this book, we come across tensors of various orders. Some examples are fluid density, velocity of a fluid particle, and stress at a point. Density is a tensor of order 0. It is completely described by $3^0 = 1$ component. Velocity of a particle is a tensor of order 1 ($n = 1$). When expressed in a coordinate system of our choice, velocity has $3^1 = 3$ scalar components. Stress at a point is a tensor of order 2. Thus, when expressed in a coordinate system, it has $3^2 = 9$ scalar components. A tensor of order zero is also called a *scalar*, and that of order 1 is also called a *vector*.

© The Author(s), under exclusive license to Springer Nature Switzerland AG 2026

S. S. Sinha, *Fundamentals of Fluid Mechanics*,
https://doi.org/10.1007/978-3-032-04210-1_1

1.1 Expressing a First-Order Tensor Using a Cartesian Coordinate System

Let us have a Cartesian coordinate system $Ox_1(\hat{e}_1)x_2(\hat{e}_2)x_3(\hat{e}_3)$, where O is the origin of the coordinate system (Fig. 1.1). The symbols $(\hat{e}_1)$, $(\hat{e}_2)$, and $(\hat{e}_3)$ are three mutually perpendicular unit vectors. The symbols x_1, x_2, and x_3 denote the coordinates of an arbitrary location (say P). If $\underline{T}$ is a tensor of order one, we express the tensor in this coordinate system as:

$$\underline{T} = T_1\hat{e}_1 + T_2\hat{e}_2 + T_3\hat{e}_3. \tag{1.1}$$

The symbols T_1, T_2, and T_3 are called the scalar components of the tensor $\underline{T}$ along the unit vectors $\hat{e}_1$, $\hat{e}_2$, and $\hat{e}_3$, respectively. Note that in this book, we represent all tensors (except scalars) using an underlined alphabetical symbol.

1.2 Transformation Rule: First-Order Tensors

Let us now consider another Cartesian coordinate system $Ox_1'(\hat{e}_1')x_2'(\hat{e}_2')x_3'(\hat{e}_3')$ (Fig. 1.2). The same tensor $\underline{T}$ can alternatively be expressed in this coordinate system as:

$$\underline{T} = T_1'\hat{e}_1' + T_2'\hat{e}_2' + T_3'\hat{e}_3'. \tag{1.2}$$

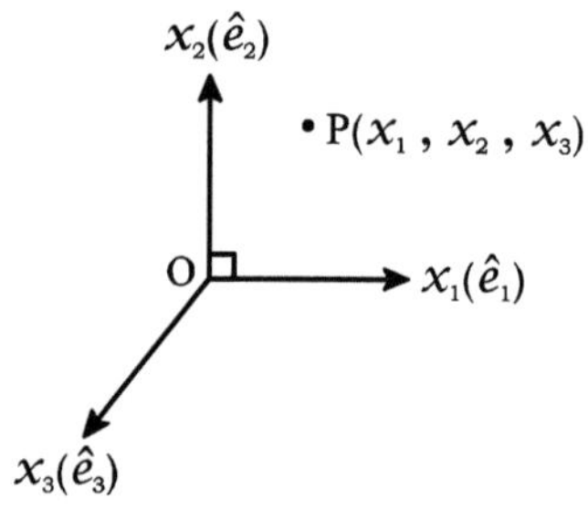

Fig. 1.1 A Cartesian coordinate system. P represents an arbitrary location with coordinates x_1, x_2, x_3

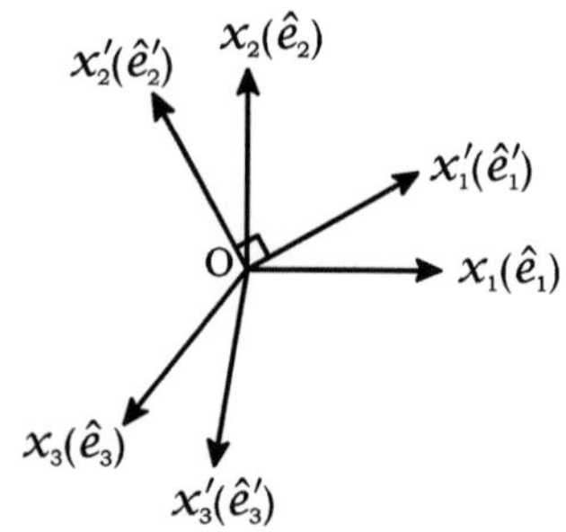

Fig. 1.2 Two different Cartesian coordinate systems

In general, $T_1^{'} \neq T_1, T_2^{'} \neq T_2$, and $T_3^{'} \neq T_3$. However, given that the tensor $\underline{T}$ must be invariant to the choice of the coordinate system used to express it, the components of the tensor in these two coordinate systems must be related. The set of relationships between these components would be called the transformation rule of the tensors of order one. To derive this transformation rule, we first express the unit vectors of the second coordinate system in terms of the unit vectors of the first coordinate system:

$$\begin{aligned} \hat{e}_1^{'} &= a_{11}\hat{e}_1 + a_{12}\hat{e}_2 + a_{13}\hat{e}_3 = \sum_{i=1}^{3} a_{1i}\hat{e}_i, \\ \hat{e}_2^{'} &= a_{21}\hat{e}_1 + a_{22}\hat{e}_2 + a_{23}\hat{e}_3 = \sum_{i=1}^{3} a_{2i}\hat{e}_i, \\ \hat{e}_3^{'} &= a_{31}\hat{e}_1 + a_{32}\hat{e}_2 + a_{33}\hat{e}_3 = \sum_{i=1}^{3} a_{3i}\hat{e}_i, \end{aligned} \tag{1.3}$$

where the symbol

$$\sum_{i=1}^{3}$$

represents the sum of three relevant terms that would be generated by allowing the index i to assume values 1, 2, and 3. The coefficients a's appearing on the right-hand side (rhs) of (1.3) are the direction cosines between the unit vectors:

$$a_{ij} = \hat{e}_i^{'} \cdot \hat{e}_j, \tag{1.4}$$

where $i \in \{1, 2, 3\}$ and $j \in \{1, 2, 3\}$. Substituting the equations for unit vectors in (1.3) of the second coordinate system in (1.2) leads to the following equation:

$$\underline{T} = T_1^{'} \sum_{i=1}^{3} a_{1i}\hat{e}_i + T_2^{'} \sum_{i=1}^{3} a_{2i}\hat{e}_i + T_3^{'} \sum_{i=1}^{3} a_{3i}\hat{e}_i. \tag{1.5}$$

Regrouping various terms on the rhs of (1.5) as coefficients of $\hat{e}_1$, $\hat{e}_2$, and $\hat{e}_3$ leads to:

$$\underline{T} = \left\{ \sum_{i=1}^{3} T_i^{'} a_{i1} \right\} \hat{e}_1 + \left\{ \sum_{i=1}^{3} T_i^{'} a_{i2} \right\} \hat{e}_2 + \left\{ \sum_{i=1}^{3} T_i^{'} a_{i3} \right\} \hat{e}_3. \tag{1.6}$$

At this point if we compare (1.6) with (1.1), we arrive at the following set of relationships between the scalar components of the tensor $\underline{T}$ in the two coordinate systems:

$$T_1 = \sum_{i=1}^{3} T_i^{'} a_{i1}, \; T_2 = \sum_{i=1}^{3} T_i^{'} a_{i2}, \text{ and } T_3 = \sum_{i=1}^{3} T_i^{'} a_{i3}. \tag{1.7}$$

This set of three equations is called the transformation rule of tensors of order one. These three relationships can be expressed in a more compact manner as:

$$T_j = \sum_{i=1}^{3} T_i' a_{ij}, \tag{1.8}$$

where $j \in \{1, 2, 3\}$ and T_j is the jth scalar component of the tensor $\underline{T}$ in the working coordinate system $Ox_1(\hat{e}_1)x_2(\hat{e}_2)x_3(\hat{e}_3)$.

1.3 Expressing a Second-Order Tensor Using a Cartesian Coordinate System

Now let us consider a second-order tensor $\underline{S}$. In a working coordinate system, it will have 9 scalar components. In the Cartesian coordinate system, $Ox_1(\hat{e}_1)x_2(\hat{e}_2)x_3(\hat{e}_3)$ this tensor is expressed as:

$$\underline{S} = S_{11}\hat{e}_1\hat{e}_1 + S_{12}\hat{e}_1\hat{e}_2 + S_{13}\hat{e}_1\hat{e}_3 + S_{21}\hat{e}_2\hat{e}_1 + S_{22}\hat{e}_2\hat{e}_2 + S_{23}\hat{e}_2\hat{e}_3 \\ + S_{31}\hat{e}_3\hat{e}_1 + S_{32}\hat{e}_3\hat{e}_2 + S_{33}\hat{e}_3\hat{e}_3 = \sum_{i=1}^{3}\sum_{j=1}^{3} S_{ij}\hat{e}_i\hat{e}_j. \tag{1.9}$$

where S_{ij} is the (ij)th scalar component of the tensor $\underline{S}$ in the working coordinate system $Ox_1(\hat{e}_1)x_2(\hat{e}_2)x_3(\hat{e}_3)$. The new mathematical entities $\hat{e}_1\hat{e}_1$, $\hat{e}_1\hat{e}_2$, $\hat{e}_1\hat{e}_3$, $\hat{e}_2\hat{e}_1$, $\hat{e}_2\hat{e}_2$, $\hat{e}_2\hat{e}_3$, $\hat{e}_3\hat{e}_1$, $\hat{e}_3\hat{e}_2$, and $\hat{e}_3\hat{e}_3$ appearing in (1.9) are called *dyads*. Each dyad is an ordered combination of two of the original coordinate system's unit vectors (equivalently, two mutually perpendicular directions). Note that $\hat{e}_1\hat{e}_2 \neq \hat{e}_2\hat{e}_1$, $\hat{e}_2\hat{e}_3 \neq \hat{e}_3\hat{e}_2$, and$\hat{e}_3\hat{e}_1 \neq \hat{e}_1\hat{e}_3$. The quantity S_{ij} is called the (ij)th component of the tensor $\underline{S}$ along the dyad $\hat{e}_i\hat{e}_j$. Using the summation symbol, S can be expressed equivalently in two ways.

$$\underline{S} = \sum_{i=1}^{3}\sum_{j=1}^{3} S_{ij}\hat{e}_i\hat{e}_j = \sum_{i=1}^{3}\sum_{j=1}^{3} S_{ji}\hat{e}_j\hat{e}_i. \tag{1.10}$$

Equation (1.10) shows that simultaneously changing the order of the indices of the dyad and that of the indices appearing with the scalar components keeps the tensor unchanged.

The *transpose* of a second-order tensor $\underline{S}$ is defined as:

$$\underline{S}^T = \sum_{i=1}^{3}\sum_{j=1}^{3} S_{ji}\hat{e}_i\hat{e}_j = \sum_{i=1}^{3}\sum_{j=1}^{3} S_{ij}\hat{e}_j\hat{e}_i. \tag{1.11}$$

where the symbol $\underline{S}^T$ is the transpose of $\underline{S}$. In general, $\underline{S}^T \neq \underline{S}$.

If a second-order tensor $\underline{S}$ is such that $\underline{S}^T = \underline{S}$, it is called a *symmetric* tensor. In contrast, a second-order tensor $\underline{S}$ is called an *antisymmetric* tensor if $\underline{S}^T = -\underline{S}$. For a symmetric tensor

$$S_{ij} = S_{ji}, \text{if } i \neq j. \tag{1.12}$$

For an antisymmetric tensor

$$S_{12} = -S_{21},\ S_{23} = -S_{32},\ S_{31} = -S_{13},\ \text{and } S_{11} = S_{22} = S_{33} = 0. \tag{1.13}$$

Any second-order tensor ($\underline{A}$) can be expressed as the sum of a symmetric tensor ($\underline{A}^{symmetric}$) and an antisymmetric tensor ($\underline{A}^{antisymmetric}$):

$$\underline{A} = \underline{A}^{symmetric} + \underline{A}^{antisymmetric}, \tag{1.14}$$

where

$$\underline{A}^{symmetric} = \frac{\underline{A} + \underline{A}^T}{2} \text{ and } \underline{A}^{antisymmetric} = \frac{\underline{A} - \underline{A}^T}{2}. \tag{1.15}$$

1.4 Transformation Rule: Second-Order Tensors

Let us now derive the transformation rule of a second-order tensor such as $\underline{S}$. Referring to the two coordinate systems of Fig. 1.2, we start with the expression of the tensor $\underline{S}$ in $Ox_1'(\hat{e}_1')x_2'(\hat{e}_2')x_3'(\hat{e}_3')$ coordinate system:

$$\underline{S} = \sum_{i=1}^{3}\sum_{j=1}^{3} S_{ij}'\hat{e}_i'\hat{e}_j'. \tag{1.16}$$

We use (1.3) to substitute the unit vectors of the second coordinate system in terms of the unit vectors of the first coordinate system. Subsequently, we separate the coefficients of the nine dyads, leading to the following expression of $\underline{S}$:

$$\underline{S} = \sum_{p=1}^{3}\sum_{q=1}^{3}\left(\sum_{i=1}^{3}\sum_{j=1}^{3} S_{ij}'a_{ip}a_{jq}\right)\hat{e}_p\hat{e}_q. \tag{1.17}$$

Earlier, we expressed tensor $\underline{S}$ directly in the first coordinate system (1.10), with summation implied over the indices i and j. However, that expression can also be expressed with summations expressed using indices p and q:

$$\underline{S} = \sum_{p=1}^{3}\sum_{q=1}^{3} S_{pq}\hat{e}_p\hat{e}_q. \tag{1.18}$$

Now comparing (1.17) with (1.18) and matching the coefficients of $\hat{e}_p\hat{e}_q$ in the two equations, we arrive at the following expression:

$$S_{pq} = \sum_{i=1}^{3}\sum_{j=1}^{3} S'_{ij}a_{ip}a_{jq}; \text{ where } p, q \in \{1, 2, 3\}. \tag{1.19}$$

The set of nine equations represented by (1.19) is the transformation rule for the second-order tensors.

1.5 Expressing Higher-Order Tensors Using a Cartesian Coordinate System

The manner of expressing the first-order and second-order tensors discussed in previous sections may be extended to higher-order tensors, as well. For an nth-order tensor with 3^n scalar components, we will require 3^n "dyad-like" members constructed with the three mutually perpendicular unit vectors of the chosen coordinate system ($\hat{e}_1$, $\hat{e}_2$, and $\hat{e}_3$). For example, a third-order tensor $\underline{Q}$ is expressed as:

$$\underline{Q} = \sum_{i=1}^{3}\sum_{j=1}^{3}\sum_{k=1}^{3} Q_{ijk}\hat{e}_i\hat{e}_j\hat{e}_k, \tag{1.20}$$

where Q_{ijk} is the (ijk)th scalar component of the tensor $\underline{Q}$ in the given Cartesian coordinate system.

1.6 Einstein's Summation Rule

So far, we have been expressing tensors using one or more summation signs $(\sum)$. While it is indeed a compact way of writing the tensors compared to writing all the 3^n terms explicitly, now onward, we wish to make the expression even more compact. We follow what is called *Einstein's summation rule*. According to this rule, the mere appearance of an index twice in a term by itself implies summation over that index. However, care must be taken that an index must not appear more than twice in any term. Following Einstein's summation rule, the final expressions of various equations (1.5)–(1.20) can be expressed as shown in (1.21)–(1.26).

An index appearing two times in a term is called a *dummy index*. In a term, the choice of an alphabet to represent a pair of dummy indices is not unique. Any symbol can be used as long as a dummy index does not appear more than twice. It is conventional to use the lowercase Latin alphabet to denote these indices. This is illustrated in some examples included in (1.21)–(1.26). In all these included examples, every index is repeated exactly two times. The order of the tensor being expressed *must* be inferred based on the number of unit vectors that stand in the sequence in the expression. The number of unit vectors defines the order of the tensor. In some cases, the number of pairs of dummy indices may coincidentally match the order of the tensor, but this is not generally true. Example 1.6 (1.26) represents the dot product of two vectors $\underline{A}$ and $\underline{B}$, which is a tensor of order zero (a scalar, no unit vectors appearing therein). However, the number of the pair of dummy indices is one.

Example 1.1

$$\underline{T} = T_1\hat{e}_1 + T_2\hat{e}_2 + T_3\hat{e}_3 = \sum_{i=1}^{3} T_i\hat{e}_i.$$

The equivalent expression using Einstein's summation rule is

$$\underline{T} = T_i\hat{e}_i = T_j\hat{e}_j. \tag{1.21}$$

Example 1.2

$$\hat{e}_1^{'} = \sum_{i=1}^{3} a_{1i}\hat{e}_i.$$

The equivalent expression using Einstein's summation rule is

$$\hat{e}_1^{'} = a_{1i}\hat{e}_i = a_{1k}\hat{e}_k. \tag{1.22}$$

Example 1.3

$$\underline{\underline{S}} = \sum_{i=1}^{3}\sum_{j=1}^{3} S_{ij}\hat{e}_i\hat{e}_j.$$

The equivalent expression using Einstein's summation rule is

$$\underline{\underline{S}} = S_{ij}\hat{e}_i\hat{e}_j = S_{iq}\hat{e}_i\hat{e}_q = S_{pq}\hat{e}_p\hat{e}_q. \tag{1.23}$$

Example 1.4

$$\underline{\underline{S}}^T = \sum_{i=1}^{3}\sum_{j=1}^{3} S_{ji}\hat{e}_i\hat{e}_j.$$

The equivalent expression using Einstein's summation rule is

$$\underline{\underline{S}}^T = S_{ji}\hat{e}_i\hat{e}_j. \tag{1.24}$$

Example 1.5

$$\underline{\underline{\underline{Q}}} = \sum_{i=1}^{3}\sum_{j=1}^{3}\sum_{k=1}^{3} Q_{ijk}\hat{e}_i\hat{e}_j\hat{e}_k.$$

The equivalent expression using Einstein's summation rule is

$$\underline{\underline{\underline{Q}}} = Q_{ijk}\hat{e}_i\hat{e}_j\hat{e}_k = Q_{ijr}\hat{e}_i\hat{e}_j\hat{e}_r = Q_{qjr}\hat{e}_q\hat{e}_j\hat{e}_r. \tag{1.25}$$

Example 1.6

$$\phi = A_1B_1 + A_2B_2 + A_3B_3 = \sum_{j=1}^{3} A_jB_j.$$

The equivalent expression using Einstein's summation rule is

$$\phi = A_jB_j. \tag{1.26}$$

Einstein's summation rule can also be used to express individual scalar components of a tensor. In (1.27) and (1.28), we include such examples (Examples 1.7 and 1.8). Note that there is already one or more pairs of dummy indices in all these expressions. Further, in each expression, there is one or more non-repeated indices. Such an index is called a *free index*. The number of free indices in an expression matches the order of the original tensor. These free indices may be assigned values 1, 2, or 3 to express various components of the tensor.

Example 1.7

$$T_j = \sum_{i=1}^{3} T_i^{'} a_{ij}.$$

The equivalent expression using Einstein's summation rule is

$$T_j = T_i^{'} a_{ij}. \tag{1.27}$$

Example 1.8

$$S_{pq} = \sum_{i=1}^{3}\sum_{j=1}^{3} S_{ij}^{'} a_{ip} a_{jq}.$$

The equivalent expression using Einstein's summation rule is

$$S_{pq} = S_{ij}^{'} a_{ip} a_{jq}. \tag{1.28}$$

In an equation, if there are free indices, all terms of the equation must have the same number of free indices, and further, those free indices must be identical in all the terms. For example, in the equation

$$S_{pq} = S_{ij}^{'} a_{ip} a_{jq}, \tag{1.29}$$

the term on the lhs and that on the rhs has exactly two free indices. Moreover, these free indices are p and q in each term. This manner of expressing tensors in a Cartesian coordinate system using free and dummy indices is called the *index notation*.

1.7 Tensor Operations

In this section, we define some tensor operations that are of relevance in our study of fluid mechanics.

1.7.1 Dot Product of Two Tensors

Let us consider two tensors $\underline{T}$ (of order $t \geq 1$) and $\underline{N}$ (of order $n \geq 1$). We define two types of dot product between the two tensors: $\underline{T} \cdot \underline{N}$ and $\underline{N} \cdot \underline{T}$. In general, these two dot products result in two different tensors. For the special case when both these tensors are of order one, then, $\underline{T} \cdot \underline{N} = \underline{N} \cdot \underline{T}$. Execution of a dot product always results in a tensor which has its order two less than the sum of the orders of the two participating tensors.

To illustrate the algebraic implementation of a dot product, we consider the dot product of a second-order tensor with a first-order tensor. We perform this illustration using the Cartesian coordinate system of Fig. 1.1 as our working coordinate system. We first express the two participating tensors in a Cartesian coordinate system and subsequently simplify the algebra as much as possible.

$$\begin{aligned}\underline{T} \cdot \underline{N} = \big[&T_{11}\hat{e}_1\hat{e}_1 + T_{12}\hat{e}_1\hat{e}_2 + T_{13}\hat{e}_1\hat{e}_3 + T_{21}\hat{e}_2\hat{e}_1 + T_{22}\hat{e}_2\hat{e}_2 + T_{23}\hat{e}_2\hat{e}_3 \\ &+ T_{31}\hat{e}_3\hat{e}_1 + T_{32}\hat{e}_3\hat{e}_2 + T_{33}\hat{e}_3\hat{e}_3\big] \cdot \big[N_1\hat{e}_1 + N_2\hat{e}_2 + N_3\hat{e}_3\big]. \end{aligned} \tag{1.30}$$

The right-hand side of equation (1.30) results into 27 individual terms.

$$\begin{aligned}\underline{T} \cdot \underline{N} = &\; T_{11}\hat{e}_1\hat{e}_1 \cdot N_1\hat{e}_1 + T_{12}\hat{e}_1\hat{e}_2 \cdot N_1\hat{e}_1 + T_{13}\hat{e}_1\hat{e}_3 \cdot N_1\hat{e}_1 \\ &+ T_{21}\hat{e}_2\hat{e}_1 \cdot N_1\hat{e}_1 + T_{22}\hat{e}_2\hat{e}_2 \cdot N_1\hat{e}_1 + T_{23}\hat{e}_2\hat{e}_3 \cdot N_1\hat{e}_1 \\ &+ T_{31}\hat{e}_3\hat{e}_1 \cdot N_1\hat{e}_1 + T_{32}\hat{e}_3\hat{e}_2 \cdot N_1\hat{e}_1 + T_{33}\hat{e}_3\hat{e}_3 \cdot N_1\hat{e}_1 \\ &+ T_{11}\hat{e}_1\hat{e}_1 \cdot N_2\hat{e}_2 + T_{12}\hat{e}_1\hat{e}_2 \cdot N_2\hat{e}_2 + T_{13}\hat{e}_1\hat{e}_3 \cdot N_2\hat{e}_2 \\ &+ T_{21}\hat{e}_2\hat{e}_1 \cdot N_2\hat{e}_2 + T_{22}\hat{e}_2\hat{e}_2 \cdot N_2\hat{e}_2 + T_{23}\hat{e}_2\hat{e}_3 \cdot N_2\hat{e}_2 \\ &+ T_{31}\hat{e}_3\hat{e}_1 \cdot N_2\hat{e}_2 + T_{32}\hat{e}_3\hat{e}_2 \cdot N_2\hat{e}_2 + T_{33}\hat{e}_3\hat{e}_3 \cdot N_2\hat{e}_2 \\ &+ T_{11}\hat{e}_1\hat{e}_1 \cdot N_3\hat{e}_3 + T_{12}\hat{e}_1\hat{e}_2 \cdot N_3\hat{e}_3 + T_{13}\hat{e}_1\hat{e}_3 \cdot N_3\hat{e}_3 \\ &+ T_{21}\hat{e}_2\hat{e}_1 \cdot N_3\hat{e}_3 + T_{22}\hat{e}_2\hat{e}_2 \cdot N_3\hat{e}_3 + T_{23}\hat{e}_2\hat{e}_3 \cdot N_3\hat{e}_3 \\ &+ T_{31}\hat{e}_3\hat{e}_1 \cdot N_3\hat{e}_3 + T_{32}\hat{e}_3\hat{e}_2 \cdot N_3\hat{e}_3 + T_{33}\hat{e}_3\hat{e}_3 \cdot N_3\hat{e}_3. \end{aligned} \tag{1.31}$$

The rule by which the dot product operates is that in an expression like $T_{12}\hat{e}_1\hat{e}_2 \cdot N_1\hat{e}_1$ of (1.31), *the right-most unit vector from the expression appearing on the left side of the dot operator should dot with the left-most unit vector from the expression appearing on the right-hand side of the dot operator*. In other words, *the "nearest" two unit vectors must dot with each other.* Thus,

$$T_{12}\hat{e}_1\hat{e}_2 \cdot N_1\hat{e}_1 = T_{12}N_1\hat{e}_1\left(\hat{e}_2 \cdot \hat{e}_1\right) = T_{12}N_1\hat{e}_1\,(0) = 0. \tag{1.32}$$

Since the three unit vectors ($\hat{e}_1$, $\hat{e}_2$, and $\hat{e}_3$) are mutually perpendicular to each other, out of the 27 individual dot products of (1.31), only those survive which involve the dot products of the same unit vectors ($\hat{e}_1 \cdot \hat{e}_1$, $\hat{e}_2 \cdot \hat{e}_2$, and $\hat{e}_3 \cdot \hat{e}_3$). Accordingly, there are nine such surviving terms.

$$\begin{aligned}\underline{T} \cdot \underline{N} = {} & T_{11}N_1\hat{e}_1 + T_{12}N_2\hat{e}_1 + T_{13}N_3\hat{e}_1 + T_{21}N_1\hat{e}_2 \\ & + T_{22}N_2\hat{e}_2 + T_{23}N_3\hat{e}_2 + T_{31}N_1\hat{e}_3 + T_{32}N_2\hat{e}_3 + T_{33}N_3\hat{e}_3.\end{aligned} \tag{1.33}$$

Using Einstein's summation rule, this dot product is written as:

$$\underline{T} \cdot \underline{N} = T_{ij}N_j\hat{e}_i. \tag{1.34}$$

Now let us consider the dot product $\underline{N} \cdot \underline{T}$. This can be expressed as:

$$\begin{aligned}\underline{N} \cdot \underline{T} = {} & \left[N_1\hat{e}_1 + N_2\hat{e}_2 + N_3\hat{e}_3\right] \cdot \left[T_{11}\hat{e}_1\hat{e}_1 + T_{12}\hat{e}_1\hat{e}_2 + T_{13}\hat{e}_1\hat{e}_3\right. \\ & \left. T_{21}\hat{e}_2\hat{e}_1 + T_{22}\hat{e}_2\hat{e}_2 + T_{23}\hat{e}_2\hat{e}_3 + T_{31}\hat{e}_3\hat{e}_1 + T_{32}\hat{e}_3\hat{e}_2 + T_{33}\hat{e}_3\hat{e}_3\right].\end{aligned} \tag{1.35}$$

Following the same rule (the nearest two unit vectors dot with each other), we arrive at:

$$\begin{aligned}\underline{N} \cdot \underline{T} = {} & N_1T_{11}\hat{e}_1 + N_1T_{12}\hat{e}_2 + N_1T_{13}\hat{e}_3 + N_2T_{21}\hat{e}_1 + N_2T_{22}\hat{e}_2 \\ & + N_2T_{23}\hat{e}_3 + N_3T_{31}\hat{e}_1 + N_3T_{32}\hat{e}_2 + N_3T_{33}\hat{e}_3.\end{aligned} \tag{1.36}$$

Using Einstein's summation rule, (1.36) is written as:

$$\underline{N} \cdot \underline{T} = N_jT_{ji}\hat{e}_i = T_{ji}N_j\hat{e}_i. \tag{1.37}$$

In the expressions of both $\underline{T} \cdot \underline{N}$ and $\underline{N} \cdot \underline{T}$ we have only one unit vector appearing on the rhs of (1.34) and (1.37). This means that both $\underline{T} \cdot \underline{N}$ and $\underline{N} \cdot \underline{T}$ are vectors (tensors of order one). Now, for the two vectors to be identical, we must compare the ith components of the two vectors. The ith component of a vector is the coefficient of $\hat{e}_i$ in its expression (1.34 or 1.37). The ith components of $\underline{T} \cdot \underline{N}$ in (1.34) and that of $\underline{N} \cdot \underline{T}$ in (1.37) are $T_{ij}N_j$ and $T_{ji}N_j$, respectively. Since $T_{ij}N_j (= T_{i1}N_1 + T_{i2}N_2 + T_{i3}N_3) \neq T_{ji}N_j (= T_{1i}N_1 + T_{2i}N_2 + T_{3i}N_3)$, we conclude that $\underline{T} \cdot \underline{N} \neq \underline{N} \cdot \underline{T}$.

At this point, we introduce a new symbol called the *Kronecker delta* (δ). This symbol has two indices as subscripts, with which we define

$$\delta_{ij} = \hat{e}_i \cdot \hat{e}_j, \tag{1.38}$$

where $\hat{e}_i$ and $\hat{e}_j$ are the ith and jth unit vectors of our Cartesian coordinate system. Clearly, (1.38) leads to:

$$\delta_{ij} = \begin{cases} 1 & \text{if } (i, j) = (1, 1), \text{ or } (2, 2), \text{ or } (3, 3) \\ 0 & \text{if } i \neq j. \end{cases} \tag{1.39}$$

Note that following the definition of δ_{ij} and simultaneously using Einstein's summation rule

$$\delta_{ii} = \delta_{11} + \delta_{22} + \delta_{33} = 3. \tag{1.40}$$

The specific purpose of introducing the Kronecker delta symbol here is to symbolically represent the dot product between the two unit vectors of the working Cartesian coordinate system. With this definition, one can avoid dealing with the expanded forms of various tensors (like what we had to do in (1.31) and (1.35)). We illustrate this by revisiting the dot product of (1.30).

$$\underline{T} \cdot \underline{N} = \left(T_{ij}\hat{e}_i\hat{e}_j\right) \cdot \left(N_p\hat{e}_p\right) = T_{ij}N_p\hat{e}_i\hat{e}_j \cdot \hat{e}_p = T_{ij}N_p\hat{e}_i\delta_{jp}. \tag{1.41}$$

Once a Kronecker delta symbol appears in a term (δ_{jp} in this particular example), we perform the following two simplifying steps: (1) remove the Kronecker delta symbol, and (2) replace either the remaining j by p or replace the remaining p by j in the term. By performing these two steps, we achieve (i) the removal of all those dot products wherein two different unit vectors participate and vanish, and (ii) retaining all those dot products wherein identical unit vectors participate. Employing these two steps, (1.41) readily simplifies as:

$$\underline{T} \cdot \underline{N} = T_{ij}N_p\hat{e}_i\delta_{jp} = T_{ij}N_j\hat{e}_i, \tag{1.42}$$

Alternatively,

$$\underline{T} \cdot \underline{N} = T_{ij}N_p\hat{e}_i\delta_{jp} = T_{ip}N_p\hat{e}_i, \tag{1.43}$$

which leads to the same outcome as what we obtained in (1.34).

1.7.2 Double Dot Product of Two Tensors

The double-dot product is defined for two tensors $\underline{\mathrm{T}}$ and $\underline{\mathrm{N}}$ of orders t and n, if $t \geq 2$ and $n \geq 2$. The resulting tensor is of order $t + n - 4$. The double-dot product operation is defined by performing two successive dot products between the right-most unit vector of the tensor on the left and the left-most unit vector of the tensor on the right. We illustrate the double-dot operation using an example wherein two second-order tensors participate:

$$\begin{aligned}\underline{T} : \underline{N} &= \left(T_{ij}\hat{e}_i\hat{e}_j\right) : \left(N_{pq}\hat{e}_p\hat{e}_q\right) = T_{ij}N_{pq}\hat{e}_i\hat{e}_j : \hat{e}_p\hat{e}_q, \\ &= T_{ij}N_{pq}\delta_{jp}\hat{e}_i \cdot \hat{e}_q = T_{ij}N_{jq}\hat{e}_i \cdot \hat{e}_q = T_{ij}N_{jq}\delta_{iq}, \\ &= T_{ij}N_{ji},\end{aligned} \tag{1.44}$$

which is a scalar (no unit vector appearing in the final expression).

1.7.3 Cross Product of Two Vectors

The cross product is defined for two vectors. The resulting quantity is also a vector. The cross product between vector $\underline{A}$ and $\underline{B}$ is defined as

$$\underline{A} \times \underline{B} = \varepsilon_{ijk} A_i B_j \hat{e}_k, \tag{1.45}$$

where the symbol ε is called the *permutation symbol* such that,

$$\varepsilon_{ijk} = \begin{cases} +1 & \text{if } i \neq j \neq k \text{ and } (i, j, k) \text{ follows a cyclic order} \\ & (1, 2, 3) \text{ or } (2, 3, 1) \text{ or } (3, 1, 2), \\ -1 & \text{if } i \neq j \neq k \text{ and } (i, j, k) \text{ follows the reversed cyclic order} \\ & (1, 3, 2) \text{ or } (3, 2, 1) \text{ or } (2, 1, 3), \\ 0 & \text{otherwise.} \end{cases} \tag{1.46}$$

Based on this definition, it follows that only when $i \neq j \neq k$, $\varepsilon_{ijk} \neq 0$. It can be verified that in case $i \neq j \neq k$, the swapping of the positions of a pair of indices changes the sign of ε_{ijk}

$$\varepsilon_{ijk} = -\varepsilon_{jik} = -(-\varepsilon_{jki}) = \varepsilon_{jki}. \tag{1.47}$$

It can be verified that the permutation symbol (1.46) leads to the following relationships

$$\begin{aligned} &\hat{e}_1 \times \hat{e}_2 = \hat{e}_3, \qquad &\hat{e}_2 \times \hat{e}_3 = \hat{e}_1, \qquad &\hat{e}_3 \times \hat{e}_1 = \hat{e}_2, \\ &\hat{e}_2 \times \hat{e}_1 = -\hat{e}_3, \qquad &\hat{e}_3 \times \hat{e}_2 = -\hat{e}_1, \qquad &\hat{e}_1 \times \hat{e}_3 = -\hat{e}_2. \end{aligned} \tag{1.48}$$

The rhs of (1.45) does lead to the following familiar expanded form of the cross product of two vectors

$$\underline{A} \times \underline{B} = (A_2 B_3 - A_3 B_2)\,\hat{e}_1 + (A_3 B_1 - A_1 B_3)\,\hat{e}_2 + (A_1 B_2 - A_2 B_1)\,\hat{e}_3. \tag{1.49}$$

1.8 The ϵ-δ Identity

It can be verified that the following relationship exists between the permutation symbol (ε) and the Kronecker delta (δ) symbols.

$$\varepsilon_{ijk}\varepsilon_{imn} = \delta_{jm}\delta_{kn} - \delta_{jn}\delta_{km}. \tag{1.50}$$

Equation (1.50) is called the ϵ-δ *identity* and often proves useful while performing algebraic manipulations of expressions involving multiple cross products. Note that the first index of the two permutation symbols on the lhs of (1.50) is identical.

1.9 Spatial Derivatives of Tensors

In later sections, when we derive the governing equations of fluid motion, we come across spatial derivatives of various kinematic and force-related quantities. Thus, we must introduce an operator ($\underline{\nabla}$) with the help of which various derivatives of space-dependent tensors can be expressed and algebraically manipulated. We refer to this operator as the *nabla* operator. Using the Cartesian coordinate system of Fig. 1.1, $\underline{\nabla}$ is expressed as:

$$\underline{\nabla} = \hat{e}_1\frac{\partial}{\partial x_1} + \hat{e}_2\frac{\partial}{\partial x_2} + \hat{e}_3\frac{\partial}{\partial x_3}, \quad \text{or } \underline{\nabla} = \hat{e}_m\frac{\partial}{\partial x_m}, \tag{1.51}$$

where $\partial/\partial x_m$ is the partial derivative operator with respect to the spatial coordinate x_m, where $m \in \{1, 2, 3\}$. In the context of this book, there are five different operations of the nabla operator which we need to understand. These are the *gradient of a tensor*, the *divergence of a tensor*, the *advection operator*, the *curl of a vector* and the *Laplacian of a tensor*,

The gradient of a tensor of order t (where $t \geq 0$) results in a tensor of order $t + 1$. We illustrate this operation using an example where $t = 2$.

$$\underline{\nabla}\,\underline{T} = \left(\hat{e}_m\frac{\partial}{\partial x_m}\right)\left(T_{ij}\hat{e}_i\hat{e}_j\right) = \hat{e}_m\frac{\partial T_{ij}}{\partial x_m}\hat{e}_i\hat{e}_j = \frac{\partial T_{ij}}{\partial x_m}\hat{e}_m\hat{e}_i\hat{e}_j. \tag{1.52}$$

Since the unit vectors $\hat{e}_1$, $\hat{e}_2$, and $\hat{e}_3$ do not depend on the spatial coordinates (x_1, x_2, and x_3), the spatial derivatives of these unit vectors do not appear in (1.52). The final expression in (1.52) has an ordered sequence of three unit vectors, which clearly shows that the resulting tensor is a third-order tensor.

The divergence of a tensor of order t is defined if $t \geq 1$. The resulting tensor is of order $t - 1$. We illustrate this operation using an example where $t = 2$.

$$\underline{\nabla}\cdot\underline{T} = \left(\hat{e}_m\frac{\partial}{\partial x_m}\right)\cdot\left(T_{ij}\hat{e}_i\hat{e}_j\right) = \frac{\partial T_{ij}}{\partial x_m}\hat{e}_m\cdot\hat{e}_i\hat{e}_j = \frac{\partial T_{ij}}{\partial x_m}\delta_{mi}\hat{e}_j,$$

$$= \frac{\partial T_{ij}}{\partial x_i} \hat{e}_j, \tag{1.53}$$

which is a first-order tensor (only one unit vector appears on the rhs of (1.53)).

The Laplacian operator (∇^2) is defined as:

$$\nabla^2 = \left(\underline{\nabla} \cdot \underline{\nabla} \right). \tag{1.54}$$

The Laplacian operator can act on a tensor of any order. The resulting tensor has the same order as the tensor on which the operator acted. In the Cartesian coordinate system, the Laplacian operator is expressed as:

$$\begin{aligned} \nabla^2 = \left(\underline{\nabla} \cdot \underline{\nabla} \right) &= \hat{e}_m \frac{\partial}{\partial x_m} \cdot \hat{e}_n \frac{\partial}{\partial x_n} = \hat{e}_m \cdot \hat{e}_n \frac{\partial^2}{\partial x_m \partial x_n} = \delta_{mn} \frac{\partial^2}{\partial x_m \partial x_n}, \\ &= \frac{\partial^2}{\partial x_m \partial x_m}. \end{aligned} \tag{1.55}$$

We illustrate the effect of this operator on a second-order tensor ($\underline{T}$):

$$\nabla^2 \underline{T} = \left(\underline{\nabla} \cdot \underline{\nabla} \right) \underline{T} = \frac{\partial^2}{\partial x_m \partial x_m} T_{ij} \hat{e}_i \hat{e}_j = \frac{\partial^2 T_{ij}}{\partial x_m \partial x_m} \hat{e}_i \hat{e}_j. \tag{1.56}$$

The right-most expression in (1.56) has an ordered pair of unit vectors ($\hat{e}_i \hat{e}_j$), which implies the resulting tensor is a second-order tensor, like $\underline{T}$, itself. However, $\nabla^2 \underline{T} \neq \underline{T}$.

The advection operator is defined as:

$$\left(\underline{V} \cdot \underline{\nabla} \right), \tag{1.57}$$

where $\underline{V}$ is a vector. The advection operator can act on a tensor of any order. The resulting tensor is of the same order as the original tensor on which the advection operator acts. Let us consider an example wherein the advection operator acts on a vector $\underline{B}$.

$$\begin{aligned} \left(\underline{V} \cdot \underline{\nabla} \right) \underline{B} &= \left(V_m \hat{e}_m \cdot \hat{e}_n \frac{\partial}{\partial x_n} \right) B_q \hat{e}_q = V_m \delta_{mn} \frac{\partial}{\partial x_n} B_q \hat{e}_q, \\ &= \left(V_n \frac{\partial}{\partial x_n} \right) B_q \hat{e}_q = V_n \frac{\partial B_q}{\partial x_n} \hat{e}_q. \end{aligned} \tag{1.58}$$

The parentheses appearing on the lhs of (1.58) imply that the dot product operation must first be performed between $\underline{V}$ and $\underline{\nabla}$, and subsequently, the resulting operator acts on $\underline{B}$.

The curl of a vector is defined as the cross product between the *nabla* operator and a vector:

$$\underline{\nabla} \times \underline{V}. \tag{1.59}$$

Expressing both the nabla operator and the vector $\underline{V}$ using our Cartesian coordinate system and (1.45), (1.59) is expressed as

$$\underline{\nabla} \times \underline{V} = \hat{e}_m \frac{\partial}{\partial x_m} \times V_p \hat{e}_p = \hat{e}_m \times \hat{e}_p \frac{\partial V_p}{\partial x_m} = \varepsilon_{mpn} \frac{\partial V_p}{\partial x_m} \hat{e}_n. \tag{1.60}$$

1.10 Index Notation and Tensor Identities

In previous sections, we introduced the index notation, primarily to enable us in expressing a tensor in the Cartesian coordinate system in a compact manner. Additionally, the index notation also proves useful in demonstrating the proofs of various tensor identities. Since tensors themselves remain invariant to the choice of the working coordinate system, so do the tensor identities. Thus, it is adequate to prove a tensor identity using merely one working coordinate system of our choice. Our Cartesian coordinate system, for which we use index notation for brevity, is indeed an apt choice to demonstrate the proof of tensor identities. For this purpose, we first use index notation to express one side (the lhs or the rhs) of the identity in the Cartesian coordinate system. Subsequently, various rules of the index notation, along with the relevant properties of the permutation symbol and the Kronecker delta symbol, are employed to simplify (and sometimes expand) the algebraic terms. Finally, these modified algebraic terms are converted back to the form which is independent of the choice of the working coordinate system. This process is illustrated using the following examples.

Example 1.9 Prove that $\underline{\nabla} \times \left(\underline{\nabla}\phi\right) = 0$.

$$\underline{\nabla} \times \left(\underline{\nabla}\phi\right) = \underline{\nabla} \times \left(\hat{e}_m \frac{\partial \phi}{\partial x_m}\right) = \varepsilon_{imk} \frac{\partial}{\partial x_i} \left(\frac{\partial \phi}{\partial x_m}\right) \hat{e}_k = \varepsilon_{imk} \frac{\partial^2 \phi}{\partial x_i \partial x_m} \hat{e}_k.$$

Now, we carefully change all i's to m's and all m's to i's in the last expression. The resulting tensor must remain unchanged, because both i and m are dummy indices. Thus,

$$\varepsilon_{imk} \frac{\partial^2 \phi}{\partial x_i \partial x_m} \hat{e}_k = \varepsilon_{mik} \frac{\partial^2 \phi}{\partial x_m \partial x_i} \hat{e}_k.$$

However, merely interchanging the positions of i and k in the permutation symbol on the rhs must reverse the sign of the tensor

$$\varepsilon_{imk} \frac{\partial^2 \phi}{\partial x_i \partial x_m} \hat{e}_k = -\varepsilon_{imk} \frac{\partial^2 \phi}{\partial x_m \partial x_i} \hat{e}_k. \tag{1.61}$$

Since $\frac{\partial^2\phi}{\partial x_m \partial x_i} = \frac{\partial^2\phi}{\partial x_i \partial x_m}$, the lhs and the rhs of (1.61) are identical except for the negative sign. For this to be true, both lhs and rhs must be zero. Thus,

$$\varepsilon_{imk} \frac{\partial^2\phi}{\partial x_i \partial x_m} \hat{e}_k = 0 \Rightarrow \underline{\nabla} \times (\underline{\nabla}\phi) = 0. \tag{1.62}$$

Example 1.10 Prove that $\underline{\nabla} \cdot (\underline{\nabla} \times \underline{V}) = 0$.

$$\begin{aligned}
\underline{\nabla} \cdot (\underline{\nabla} \times \underline{V}) &= \hat{e}_m \frac{\partial}{\partial x_m} \cdot \left(\hat{e}_p \frac{\partial}{\partial x_p} \times V_q \hat{e}_q \right) \\
&= \hat{e}_m \frac{\partial}{\partial x_m} \cdot \left(\varepsilon_{pqr} \frac{\partial V_q}{\partial x_p} \hat{e}_r \right) = \varepsilon_{pqr} \frac{\partial^2 V_q}{\partial x_m \partial x_p} \hat{e}_m \cdot \hat{e}_r, \\
&= \varepsilon_{pqr} \frac{\partial^2 V_q}{\partial x_m \partial x_p} \delta_{mr} = \varepsilon_{pqm} \frac{\partial^2 V_q}{\partial x_m \partial x_p}.
\end{aligned} \tag{1.63}$$

Now, we carefully change all p's to m's and all m's to p's in the last expression. The resulting tensor must remain unchanged. Thus,

$$\varepsilon_{pqm} \frac{\partial^2 V_q}{\partial x_m \partial x_p} = \varepsilon_{mqp} \frac{\partial^2 V_q}{\partial x_p \partial x_m}. \tag{1.64}$$

However, merely interchanging the positions of m and p in the permutation symbol on the rhs must reverse the sign of the tensor leading to

$$\varepsilon_{pqm} \frac{\partial^2 V_q}{\partial x_m \partial x_p} = -\varepsilon_{pqm} \frac{\partial^2 V_q}{\partial x_p \partial x_m}. \tag{1.65}$$

We observe that the lhs and the rhs of (1.65) are identical except for the negative sign. For this to be true, both lhs and rhs must be zero.

$$\varepsilon_{pqm} \frac{\partial^2 V_q}{\partial x_p \partial x_m} = 0 \Rightarrow \underline{\nabla} \cdot (\underline{\nabla} \times \underline{V}) = 0. \tag{1.66}$$

Example 1.11 Prove that

$$\underline{\nabla} \times (\underline{\omega} \times \underline{V}) = (\underline{V} \cdot \underline{\nabla})\, \underline{\omega} - (\underline{\nabla} \cdot \underline{\omega})\, \underline{V} + (\underline{\nabla} \cdot \underline{V})\, \underline{\omega} - (\underline{\omega} \cdot \underline{\nabla})\, \underline{V}.$$

$$\begin{aligned}
\underline{\nabla} \times (\underline{\omega} \times \underline{V}) &= \underline{\nabla} \times \left(\varepsilon_{ijk} \omega_i V_j \hat{e}_k \right) = \epsilon_{pkr} \frac{\partial}{\partial x_p} \left(\varepsilon_{ijk} \omega_i V_j \right) \hat{e}_r, \\
&= \varepsilon_{pkr} \varepsilon_{ijk} \frac{\partial \omega_i}{\partial x_p} V_j \hat{e}_r + \varepsilon_{pkr} \varepsilon_{ijk} \frac{\partial V_j}{\partial x_p} \omega_i \hat{e}_r.
\end{aligned} \tag{1.67}$$

Both the terms in the rhs of the last expression in (1.67) has two permutation symbols with one common index. The $\epsilon - \delta$ identity (1.50) can be applied here. However, before we do so, we must re-arrange the indices of both the permutation symbols such that common index appears as the first index of each of these symbols.

$$
\begin{aligned}
&\varepsilon_{pkr}\varepsilon_{ijk}\frac{\partial \omega_i}{\partial x_p}V_j\hat{e}_r + \varepsilon_{pkr}\varepsilon_{ijk}\frac{\partial V_j}{\partial x_p}\omega_i\hat{e}_r \\
&\quad = (-\varepsilon_{kpr})(-\varepsilon_{kji})\frac{\partial \omega_i}{\partial x_p}V_j\hat{e}_r + (-\varepsilon_{kpr})(-\varepsilon_{kji})\frac{\partial V_j}{\partial x_p}\omega_i\hat{e}_r, \\
&\quad = \varepsilon_{kpr}\varepsilon_{kji}\frac{\partial \omega_i}{\partial x_p}V_j\hat{e}_r + \varepsilon_{kpr}\varepsilon_{kji}\frac{\partial V_j}{\partial x_p}\omega_i\hat{e}_r, \\
&\quad = \left[\delta_{pj}\delta_{ri}\frac{\partial \omega_i}{\partial x_p}V_j\hat{e}_r - \delta_{pi}\delta_{rj}\frac{\partial \omega_i}{\partial x_p}V_j\hat{e}_r\right] \\
&\qquad + \left[\delta_{pj}\delta_{ri}\frac{\partial V_j}{\partial x_p}\omega_i\hat{e}_r - \delta_{pi}\delta_{rj}\frac{\partial V_j}{\partial x_p}\omega_i\hat{e}_r\right], \\
&\quad = \left[V_p\frac{\partial \omega_r}{\partial x_p}\hat{e}_r - \frac{\partial \omega_p}{\partial x_p}V_r\hat{e}_r\right] + \left[\frac{\partial V_p}{\partial x_p}\omega_r\hat{e}_r - \omega_p\frac{\partial V_r}{\partial x_p}\hat{e}_r\right], \\
&\quad = \left[\left(\underline{V}\cdot\underline{\nabla}\right)\underline{\omega} - \left(\underline{\nabla}\cdot\underline{\omega}\right)\underline{V}\right] + \left[\left(\underline{\nabla}\cdot\underline{V}\right)\underline{\omega} - \left(\underline{\omega}\cdot\underline{\nabla}\right)\underline{V}\right], \\
&\Rightarrow \underline{\nabla}\times\left(\underline{\omega}\times\underline{V}\right) = \left(\underline{V}\cdot\underline{\nabla}\right)\underline{\omega} - \left(\underline{\nabla}\cdot\underline{\omega}\right)\underline{V} + \left(\underline{\nabla}\cdot\underline{V}\right)\underline{\omega} - \left(\underline{\omega}\cdot\underline{\nabla}\right)\underline{V}.
\end{aligned}
\tag{1.68}
$$

Example 1.12 Prove that $\underline{V}\times\left(\underline{\nabla}\times\underline{V}\right) = \frac{1}{2}\underline{\nabla}\left(\underline{V}\cdot\underline{V}\right) - \left(\underline{V}\cdot\underline{\nabla}\right)\underline{V}$.

$$
\begin{aligned}
\underline{V}\times\left(\underline{\nabla}\times\underline{V}\right) &= V_p\hat{e}_p\times\left(\hat{e}_i\frac{\partial}{\partial x_i}\times V_j\hat{e}_j\right) = V_p\hat{e}_p\times\left(\varepsilon_{ijk}\frac{\partial V_j}{\partial x_i}\hat{e}_k\right), \\
&= \varepsilon_{pkr}V_p\varepsilon_{ijk}\frac{\partial V_j}{\partial x_i}\hat{e}_r = \left(-\varepsilon_{kpr}\right)V_p\left(-\varepsilon_{kji}\right)\frac{\partial V_j}{\partial x_i}\hat{e}_r, \\
&= \varepsilon_{kpr}\varepsilon_{kji}V_p\frac{\partial V_j}{\partial x_i}\hat{e}_r = \left(\delta_{pj}\delta_{ri} - \delta_{pi}\delta_{rj}\right)V_p\frac{\partial V_j}{\partial x_i}\hat{e}_r, \\
&= \delta_{pj}\delta_{ri}V_p\frac{\partial V_j}{\partial x_i}\hat{e}_r - \delta_{pi}\delta_{rj}V_p\frac{\partial V_j}{\partial x_i}\hat{e}_r, \\
&= \delta_{ri}V_j\frac{\partial V_j}{\partial x_i}\hat{e}_r - \delta_{rj}V_i\frac{\partial V_j}{\partial x_i}\hat{e}_r = V_j\frac{\partial V_j}{\partial x_i}\hat{e}_i - V_i\frac{\partial V_j}{\partial x_i}\hat{e}_j, \\
&= \hat{e}_i\frac{1}{2}\frac{\partial\left(V_jV_j\right)}{\partial x_i} - V_i\frac{\partial V_j}{\partial x_i}\hat{e}_j, \\
&= \underline{\nabla}\left[\frac{\left(\underline{V}\cdot\underline{V}\right)}{2}\right] - \left(\underline{V}\cdot\underline{\nabla}\right)\underline{V}, \\
\Rightarrow \underline{V}\times\left(\underline{\nabla}\times\underline{V}\right) &= \frac{\underline{\nabla}\left(\underline{V}\cdot\underline{V}\right)}{2} - \left(\underline{V}\cdot\underline{\nabla}\right)\underline{V}.
\end{aligned}
\tag{1.69}
$$

Example 1.13 Prove that $\left[(\underline{\nabla}\phi)\cdot\underline{\nabla}\right]\underline{\nabla}\phi = \frac{1}{2}\underline{\nabla}\left[(\underline{\nabla}\phi)\cdot(\underline{\nabla}\phi)\right]$.

$$\begin{aligned}
\left[(\underline{\nabla}\phi)\cdot\underline{\nabla}\right]\underline{\nabla}\phi &= \left[\left(\hat{e}_m\frac{\partial}{\partial x_m}\phi\right)\cdot\hat{e}_n\frac{\partial}{\partial x_n}\right]\hat{e}_p\frac{\partial\phi}{\partial x_p},\\
&= \hat{e}_m\cdot\hat{e}_n\frac{\partial\phi}{\partial x_m}\frac{\partial}{\partial x_n}\left(\hat{e}_p\frac{\partial\phi}{\partial x_p}\right) = \delta_{mn}\frac{\partial\phi}{\partial x_m}\frac{\partial^2\phi}{\partial x_n\partial x_p}\hat{e}_p,\\
&= \hat{e}_p\frac{\partial\phi}{\partial x_m}\frac{\partial^2\phi}{\partial x_m\partial x_p} = \hat{e}_p\frac{\partial\phi}{\partial x_m}\frac{\partial}{\partial x_p}\left(\frac{\partial\phi}{\partial x_m}\right),\\
&= \hat{e}_p\frac{1}{2}\frac{\partial}{\partial x_p}\left(\frac{\partial\phi}{\partial x_m}\frac{\partial\phi}{\partial x_m}\right) = \frac{1}{2}\underline{\nabla}\left[(\underline{\nabla}\phi)\cdot(\underline{\nabla}\phi)\right],\\
\Rightarrow \left[(\underline{\nabla}\phi)\cdot\underline{\nabla}\right]\underline{\nabla}\phi &= \frac{1}{2}\underline{\nabla}\left[(\underline{\nabla}\phi)\cdot(\underline{\nabla}\phi)\right].
\end{aligned} \tag{1.70}$$

Example 1.14 Prove that $\underline{S}:\underline{W}=0$ where $\underline{S}$ and $\underline{W}$ represent a symmetric and an antisymmetric second-order tensor, respectively.

$$\begin{aligned}
\underline{S}:\underline{W} &= S_{ij}\hat{e}_i\hat{e}_j : W_{mn}\hat{e}_m\hat{e}_n = S_{ij}\hat{e}_i\cdot W_{mn}\delta_{jm}\hat{e}_n,\\
&= S_{ij}\hat{e}_i\cdot W_{jn}\hat{e}_n = S_{ij}W_{jn}\delta_{in} = S_{ij}W_{ji}.
\end{aligned} \tag{1.71}$$

The quantity on the rhs of (1.71) must remain unchanged if all i's are made j's, and all j's are made i's. Thus,

$$\underline{S}:\underline{W} = S_{ij}W_{ji} = S_{ji}W_{ij}. \tag{1.72}$$

Since $\underline{S}$ is a symmetric tensor $S_{ji}=S_{ij}$. Thus,

$$S_{ji}W_{ij} = S_{ij}W_{ij}. \tag{1.73}$$

Since $\underline{W}$ is an antisymmetric tensor $W_{ij}=-W_{ji}$. Thus,

$$S_{ij}W_{ij} = -S_{ij}W_{ji}. \tag{1.74}$$

Now equating the rhs of (1.74) directly to the right-most term of (1.71), we conclude

$$-S_{ij}W_{ji} = S_{ij}W_{ji}. \tag{1.75}$$

Equation (1.75) has both its sides identical except for the negative sign. This implies,

$$S_{ij}W_{ji} = 0 \Rightarrow \underline{S}:\underline{W}=0. \tag{1.76}$$

In the subsequent chapters of this book, we frequently refer to these useful identities (Examples 1.9–1.14) while deriving the governing equations of fluid motion.

1.11 Expressing Tensors Using the Cylindrical-Polar Coordinate System

In some scenarios, it may be easier to express tensors and perform algebraic manipulations on them using a coordinate system other than a Cartesian coordinate system. The cylindrical-polar coordinate system is one such alternative. To describe a cylindrical-polar coordinate system, we do need a frame-fixed Cartesian coordinate system in the background. A cylindrical coordinate system uses three mutually perpendicular unit vectors $\hat{e}_r$, $\hat{e}_\theta$, and $\hat{e}_z$. The $\hat{e}_z$ unit vector is identical to the $\hat{e}_3$ unit vector of the background Cartesian coordinate system. These unit vectors are shown in Fig. 1.3. In this figure, P represents an arbitrary location in space. The projection of the vector pointing from the origin O to this location P on the plane of $x_1(\hat{e}_1) - x_2(\hat{e}_2)$ is shown as a segment OP'. The unit vector $\hat{e}_r$ is oriented along this line segment. Thus, the vector $\hat{e}_r$ always lies in the $x_1(\hat{e}_1) - x_2(\hat{e}_2)$ plane. The second unit vector $\hat{e}_\theta$ also lies in the same plane but is normal to $\hat{e}_r$. The direction of $\hat{e}_\theta$ is such that $\hat{e}_r \times \hat{e}_\theta = \hat{e}_z$. The symbols r, θ, and z are the coordinates of location P. The symbol r equals the length of the segment of OP' ($r \geq 0$). The symbol θ represents the angle subtended by the segment OP' on the $x_1(\hat{e}_1)$ axis ($\theta \geq 0$). The sense of increasing θ is determined as shown in Fig. 1.3. The third coordinate z is the scalar component of the position vector of point P along the $\hat{e}_z$ unit vector ($-\infty < z < \infty$). The orientations of the unit vectors $\hat{e}_r$ and $\hat{e}_\theta$ are not constants but vary when θ is changed. This differs from the Cartesian coordinate system, wherein none of the unit vectors change with any spatial coordinates (x_1, x_2, or x_3). Using the cylindrical-polar coordinate system of Fig. 1.3, the position vector of P ($\underline{r}_{PO}$) is expressed as:

$$\underline{r}_{PO} = \underline{r}_{P'O} + \underline{r}_{PP'} = r\hat{e}_r + z\hat{e}_z. \tag{1.77}$$

An arbitrary vector $\underline{A}$ is expressed in the cylindrical-polar coordinate system as:

$$\underline{A} = A_r\hat{e}_r + A_\theta\hat{e}_\theta + A_z\hat{e}_z, \tag{1.78}$$

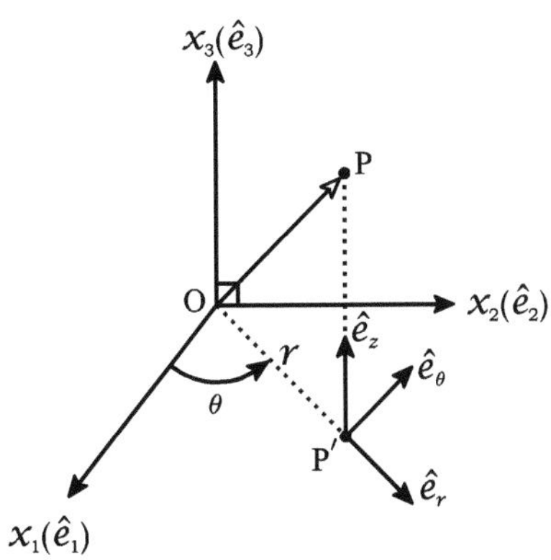

Fig. 1.3 The cylindrical-polar coordinate system

where A_r, A_θ, and A_z are called the scalar components of the vector $\underline{A}$ in this coordinate system. To express tensors of higher order, we must construct an ordered sequence of unit vectors $\hat{e}_r$, $\hat{e}_\theta$, and $\hat{e}_z$. For example, the expression of a second-order tensor $\underline{T}$ in the cylindrical-polar coordinate system is:

$$\underline{T} = T_{rr}\hat{e}_r\hat{e}_r + T_{r\theta}\hat{e}_r\hat{e}_\theta + T_{rz}\hat{e}_r\hat{e}_z + T_{\theta r}\hat{e}_\theta\hat{e}_r + T_{\theta\theta}\hat{e}_\theta\hat{e}_\theta + T_{\theta z}\hat{e}_\theta\hat{e}_z + T_{zr}\hat{e}_z\hat{e}_r + T_{z\theta}\hat{e}_z\hat{e}_\theta + T_{zz}\hat{e}_z\hat{e}_z. \quad (1.79)$$

Using the definition of the $\underline{\nabla}$ operator (1.51), the corresponding expression in the cylindrical-polar coordinate system can be derived. For this derivation, referring to Fig. 1.3, we first express the coordinates of the cylindrical-polar coordinate system (r, θ, and z) in terms of those of the Cartesian coordinate system (x_1, x_2, and x_3).

$$r = \sqrt{x_1^2 + x_2^2},\ \theta = tan^{-1}\left(\frac{x_2}{x_1}\right),\ z = x_3. \quad (1.80)$$

Subsequently, we express the partial derivatives with respect to the individual coordinates of the Cartesian coordinate system in terms of the partial derivatives with respect to the individual coordinates of the cylindrical-polar coordinate system. The appropriate chain rules of partial derivatives are employed.

$$\begin{aligned} \frac{\partial}{\partial x_1} &= \frac{\partial r}{\partial x_1}\frac{\partial}{\partial r} + \frac{\partial \theta}{\partial x_1}\frac{\partial}{\partial \theta} + \frac{\partial z}{\partial x_1}\frac{\partial}{\partial z}, \\ \frac{\partial}{\partial x_2} &= \frac{\partial r}{\partial x_2}\frac{\partial}{\partial r} + \frac{\partial \theta}{\partial x_2}\frac{\partial}{\partial \theta} + \frac{\partial z}{\partial x_2}\frac{\partial}{\partial z}, \\ \frac{\partial}{\partial x_3} &= \frac{\partial}{\partial z}. \end{aligned} \quad (1.81)$$

Using the relationship between the Cartesian and the cylindrical-polar coordinates (1.80) in (1.81) we arrive at the following

$$\begin{aligned} \frac{\partial}{\partial x_1} &= \cos\theta\frac{\partial}{\partial r} - \frac{\sin\theta}{r}\frac{\partial}{\partial \theta}, \\ \frac{\partial}{\partial x_2} &= \sin\theta\frac{\partial}{\partial r} + \frac{\cos\theta}{r}\frac{\partial}{\partial \theta}. \end{aligned} \quad (1.82)$$

Further, we express the unit vectors of the Cartesian coordinate system in terms of those of the cylindrical-polar coordinate system

$$\hat{e}_1 = \hat{e}_r\cos\theta - \hat{e}_\theta\sin\theta,\ \hat{e}_2 = \hat{e}_r\sin\theta + \hat{e}_\theta\cos\theta,\ \hat{e}_3 = \hat{e}_z. \quad (1.83)$$

Finally, using (1.82) and (1.83) to make appropriate substitutions in (1.51) leads to the expression of $\underline{\nabla}$ in the cylindrical-polar coordinate system

$$\underline{\nabla} = \hat{e}_r \frac{\partial}{\partial r} + \hat{e}_\theta \frac{1}{r} \frac{\partial}{\partial \theta} + \hat{e}_z \frac{\partial}{\partial z}. \tag{1.84}$$

Since the orientations of the unit vectors $\hat{e}_r$ and $\hat{e}_\theta$ change with θ, care must be taken to appropriately include the non-zero contributions of the partial derivatives of these unit vectors with respect to the spatial coordinate θ. These non-zero contributions arise while dealing with the $\underline{\nabla}$ operator (the gradient, divergence, Laplacian, advection, and curl operations). With reference to Fig. 1.3, derivatives can be ascertained by first expressing the $\hat{e}_r$ and $\hat{e}_\theta$ unit vectors in the Cartesian coordinate system:

$$\hat{e}_r = \cos\theta \hat{e}_1 + \sin\theta \hat{e}_2 \text{ and } \hat{e}_\theta = -\sin\theta \hat{e}_1 + \cos\theta \hat{e}_2. \tag{1.85}$$

Taking the partial derivatives of these expressions with respect to θ leads to:

$$\frac{\partial \hat{e}_r}{\partial \theta} = -\sin\theta \hat{e}_1 + \cos\theta \hat{e}_2 \text{ and } \frac{\partial \hat{e}_\theta}{\partial \theta} = -\cos\theta \hat{e}_1 - \sin\theta \hat{e}_2. \tag{1.86}$$

Comparing (1.85) and (1.86), we conclude:

$$\frac{\partial \hat{e}_r}{\partial \theta} = \hat{e}_\theta \text{ and } \frac{\partial \hat{e}_\theta}{\partial \theta} = -\hat{e}_r. \tag{1.87}$$

Except for these two partial derivatives, the other seven partial derivatives of various unit vectors with respect to the coordinates of the cylindrical-polar coordinate system do vanish:

$$\begin{aligned} &\frac{\partial \hat{e}_r}{\partial r} = 0, \quad \frac{\partial \hat{e}_r}{\partial z} = 0, \\ &\frac{\partial \hat{e}_\theta}{\partial r} = 0, \quad \frac{\partial \hat{e}_\theta}{\partial z} = 0, \\ &\frac{\partial \hat{e}_z}{\partial r} = 0, \quad \frac{\partial \hat{e}_z}{\partial \theta} = 0, \quad \frac{\partial \hat{e}_z}{\partial z} = 0. \end{aligned} \tag{1.88}$$

It can be verified that the expressions of various operators when expressed in the cylindrical-polar coordinate system are

$$\underline{\nabla} \cdot \underline{\nabla} = \frac{1}{r}\left[\frac{\partial}{\partial r}\left(r\frac{\partial}{\partial r}\right) + \frac{1}{r}\frac{\partial^2}{\partial \theta^2} + r\frac{\partial^2}{\partial z^2}\right], \tag{1.89}$$

$$\underline{\nabla} \cdot \underline{V} = \frac{1}{r}\left[\frac{\partial (rV_r)}{\partial r} + \frac{\partial V_\theta}{\partial \theta} + r\frac{\partial V_z}{\partial z}\right], \tag{1.90}$$

$$\begin{aligned} \left(\underline{V} \cdot \underline{\nabla}\right) \underline{V} = {} & \hat{e}_r \left(V_r \frac{\partial V_r}{\partial r} + \frac{V_\theta}{r}\frac{\partial V_r}{\partial \theta} - \frac{V_\theta^2}{r} + V_z \frac{\partial V_r}{\partial z}\right) \\ & + \hat{e}_\theta \left(\frac{V_\theta V_r}{r} + V_r \frac{\partial V_\theta}{\partial r} + \frac{V_\theta}{r}\frac{\partial V_\theta}{\partial \theta} + V_z \frac{\partial V_\theta}{\partial z}\right) \\ & + \hat{e}_z \left(V_r \frac{\partial V_z}{\partial r} + \frac{V_\theta}{r}\frac{\partial V_z}{\partial \theta} + V_z \frac{\partial V_z}{\partial z}\right). \end{aligned} \tag{1.91}$$

Chapter 2
Mechanics: A Brief Review of the Fundamentals

Our study of fluid mechanics is based on the fundamental concepts of *particle mechanics*. We briefly review these in this chapter. The reader is referred to [1] for more details.

1. A *particle* is defined to be a point mass.
2. A *body* is defined as a continuous distribution of particles with some inter-particle binding forces.
3. A *rigid body* is defined as a body wherein all inter-particle distances remain unchanged at all times.
4. A *deformable* body is a non-rigid body wherein some or all inter-particle distances may change.
5. A *reference frame* is a real or an imaginary rigid body in contrast to which or *with respect to* which the motion of a particle or a system of particles is observed by an observer. In this book, a reference frame will be symbolically represented as shown in Fig. 2.1, which depicts a rigid body with a human observer who is fixed to this body itself while he observes the motion of the particles/other bodies present in his surroundings. We use a capitalized Latin alphabet to label a reference frame (say F).

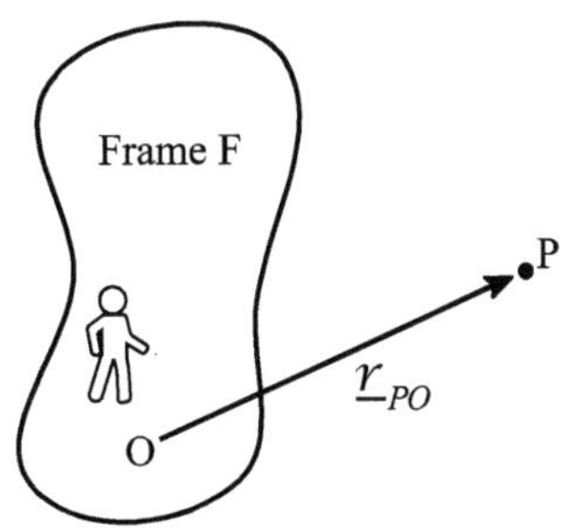

Fig. 2.1 Symbolic representation of a reference frame F. O is a point fixed to the reference frame. $\underline{r}_{PO}$ is the position vector of a particle P

© The Author(s), under exclusive license to Springer Nature Switzerland AG 2026
S. S. Sinha, *Fundamentals of Fluid Mechanics*,
https://doi.org/10.1007/978-3-032-04210-1_2

6. *Origin* of a reference frame is a point fixed to the reference frame and is used to measure the relative displacement of a particle of interest.
7. The *position vector* of a particle P measured from the origin (say O) of the reference frame is a vector with its head placed at the instantaneous location of P and its tail being at point O. Symbolically, we represent the position vector of P as $\underline{r}_{PO}$ (Fig. 2.1).
8. The rate of change of a tensor $\underline{T}$ (order $\geq$ 1) with time (t) and with respect to reference frame F is defined as:

$$\left.\frac{d\underline{T}}{dt}\right|_F = \lim_{\Delta t \to 0} \frac{\underline{T}(t + \Delta t) - \underline{T}(t)}{\Delta t}. \tag{2.1}$$

where Δt represents a small increment in time, and $\underline{T}(t)$ represents the state of the tensor at time t. The subscript F on the left-hand side of (2.1) signifies that, in general, the rate of change of a tensor (order $\geq$ 1) is a frame-dependent quantity. If the same tensor $\underline{T}$ is being simultaneously observed with respect to another reference frame, say G, then, in general,

$$\left.\frac{d\underline{T}}{dt}\right|_F \neq \left.\frac{d\underline{T}}{dt}\right|_G. \tag{2.2}$$

In contrast, the rate of change of a given scalar, say S, is always frame-independent:

$$\left.\frac{dS}{dt}\right|_F = \left.\frac{dS}{dt}\right|_G = \frac{dS}{dt}. \tag{2.3}$$

Since the derivative is frame-independent, there is no requirement to include the symbols F or G as a subscript in the symbol $\frac{dS}{dt}$.
9. *Kinematics* is the study of the motion of a particle or a system (collection) of particles without any reference to the cause of that motion.
10. *Dynamics* is the study of the motion of a particle or a system of particles and its relationship with the forces that cause that motion.
11. *Statics* is a study that identifies and examines the influence of those forces which keep a particle or system of particles at rest.
12. The *velocity* vector of a particle P with respect to a frame F is defined as:

$$\underline{V}_{P|F} = \lim_{\Delta t \to 0} \frac{\underline{r}_{PO}(t + \Delta t) - \underline{r}_{PO}(t)}{\Delta t}, \tag{2.4}$$

where $\underline{r}_{PO}(t)$ represents the position vector of the particle P at time t (Fig. 2.2) measured from point O, which must be a point fixed to the reference frame F (origin of the reference frame). It can be verified that if the observer chooses another point O' as the origin of the reference frame, $\underline{V}_{p|F}$ still turns out to be the same vector when O is chosen as the origin (Fig. 2.2).

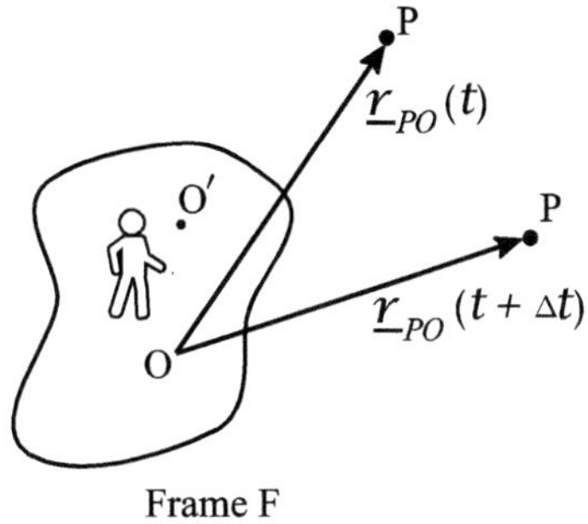

Fig. 2.2 Position vectors of the particle P at two different time instants

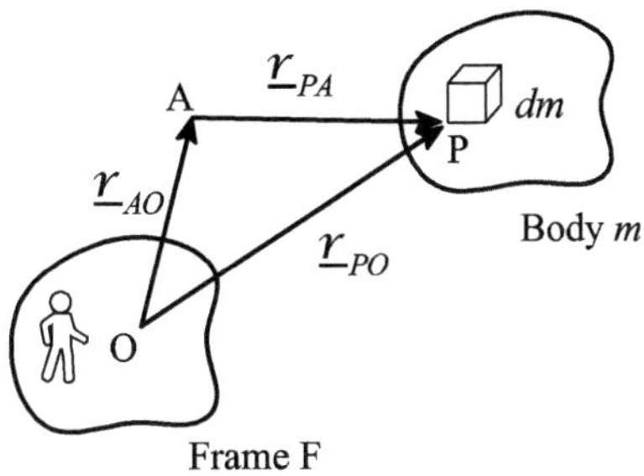

Fig. 2.3 Determination of angular momentum of body m with respect to frame F about point A where dm is the mass of an infinitesimal part of the body m and P is the location of the infinitesimal mass element on the body

$$\begin{aligned}\underline{V}_{P|F} &= \lim_{\Delta t\to 0}\frac{\underline{r}_{PO}(t+\Delta t)-\underline{r}_{PO}(t)}{\Delta t},\\ &= \lim_{\Delta t\to 0}\frac{\underline{r}_{PO'}(t+\Delta t)-\underline{r}_{PO'}(t)}{\Delta t},\end{aligned} \tag{2.5}$$

where O' is another point which is fixed to the reference frame F.

13. The *velocity of particle P relative to another particle Q with respect to frame F* is defined as:

$$\underline{V}_{PQ|F} = \underline{V}_{P|F} - \underline{V}_{Q|F}. \tag{2.6}$$

14. The *acceleration* vector of particle P with respect to reference frame F is defined as:

$$\underline{a}_{P|F} = \lim_{\Delta t\to 0}\frac{\underline{V}_{P|F}(t+\Delta t)-\underline{V}_{P|F}(t)}{\Delta t}. \tag{2.7}$$

If the motion of particle P is being simultaneously observed with respect to two different reference frames, in general, $\underline{V}_{P|F} \neq \underline{V}_{P|G}$ and $\underline{a}_{P|F} \neq \underline{a}_{P|G}$.

15. The *momentum vector* of a body m with respect to frame F is defined as:

$$\underline{p}_{|F} = \int_m \underline{V}_{P|F}\,dm, \tag{2.8}$$

where dm represents the mass of an infinitesimal element of the body, and $V_{P|F}$ is the velocity of this infinitesimal element (Fig. 2.3). The integration in (2.8) is to be performed over the entire body.

16. The *angular momentum vector of a body about point A and with respect to frame F* is defined as:

$$\underline{H}_{A|F} = \int_m \underline{r}_{PA} \times \underline{V}_{P|F} dm. \tag{2.9}$$

where $\underline{r}_{PA}$ represents the position vector of the infinitesimal mass element dm measured from point A (Fig. 2.3). The integration in (2.9) is to be performed over the entire body of mass m.

17. *Moment due to a force $\underline{F}$* about a point O is defined as $\underline{r}_{PO} \times \underline{F}$, where P is the point where the force vector is being applied. This moment is independent of point P as long as P lies on the line of action of the force vector.
18. *A couple of forces* or a *couple* is a set of two forces that are equal and opposite but have different lines of action. The algebraic sum of these two forces is zero. However, the algebraic sum of the moments due to a couple of forces about an arbitrary point is non-zero and is actually identical about all points in space. The moment due to a couple of forces is also called a *pure moment.*
19. *A free body diagram* is a schematic showing all the external forces and their points of application on an identified body of interest.
20. *A system of forces* is the set of all external forces that are acting on a body of interest. In general, this system of forces can cause changes in both the momentum and the angular momentum of the body.
21. *Euler's first axiom.* The vector sum of all the external forces acting on a body equals the rate of change of momentum of the body with respect to an inertial reference frame.
22. *Euler's second axiom.* The vector sum of the moment of all the external forces acting on a body about a point (say A), which is fixed in an inertial reference frame, equals the rate of change of angular momentum of the body about the same point A. It follows from Euler's second axiom that the vector sum of the moment of all the external forces acting on a body about its center of mass equals the rate of change of angular momentum of the body calculated about its center of mass.
23. *Equivalent force systems.* Two different systems of forces are said to be equivalent if (i) the vector sums of the constituent forces of the two systems are identical, and (ii) the vector sums of the moments of the constituent forces from the two systems about any one (arbitrarily chosen) point is identical.
24. *The resultant force system at a point* of a given force system consists of a single force vector acting on the body at that point and an associated pure moment vector, such that the original force system and the resultant force systems are equivalent to each other. For some special force systems, there may exist a set of points in space at which the resultant force system consists of just one single force vector and no associated pure moment.

Chapter 3
Description of Fluid Motion

The word *fluids* refers to gases and liquids. Fluids are comprised of molecules. One way to describe the motion of a fluid is to directly track the motion of all the constituent molecules, employing Euler's first axiom. Even though this approach sounds simple, it may prove to be impractical to implement because, typically, a small chunk of fluid comprises a very large number of molecules.

3.1 The Continuum Description

For many engineering flows of interest, it is preferred to describe the motion of fluid using what is called the *continuum* description. The continuum description does not track the motion of individual molecules but tracks the motion of individual *fluid particles*. A fluid particle is assumed to be a point mass in the continuum description and is characterized by its velocity, acceleration (with respect to a reference frame in context), density, pressure, and temperature. We refer to these quantities as the properties of a fluid particle. Ascertaining the values of these properties for each fluid particle in a domain of interest culminates in describing the motion of the fluid in that domain.

Even though the continuum description does not directly reference the constituent molecules, it is insightful to be aware of the relationships between the molecular motion and the properties of a fluid particle. Various properties associated with a fluid particle are indeed certain manifestations of the averaged motion of the numerous molecules that constitute a fluid particle. We must acknowledge that even though a fluid particle is assumed to be a point mass in the continuum description, on the absolute length scale, a fluid particle does have small but finite dimensions and does occupy a small but finite volume containing numerous molecules of the fluid substance. The absolute dimensions of a fluid particle can be deemed comparable to the dimensions of the surface of a small sensor with which we measure velocity/temperature/pressure in a fluid. Thus, it is typical to visualize a fluid particle at

© The Author(s), under exclusive license to Springer Nature Switzerland AG 2026
S. S. Sinha, *Fundamentals of Fluid Mechanics*,
https://doi.org/10.1007/978-3-032-04210-1_3

While describing fluid motion, it is of interest to inquire about the rate of change of a dependent variable of a fluid particle. Let us examine the partial derivative of the dependent variables of the Lagrangian description with respect to time. At time t, the partial derivative of a Lagrangian dependent variable $\phi^+(t, \underline{Y})$ with respect to time is as:

$$\frac{\partial \phi^+(t, \underline{Y})}{\partial t} = \lim_{\Delta t \to 0} \frac{\phi^+(t + \Delta t, \underline{Y}) - \phi^+(t, \underline{Y})}{\Delta t}. \tag{3.2}$$

On the right-hand side, two observations of the dependent variable are being used. Since the derivative of the dependent variable is partial with respect to t, by the definition of a partial derivative, both observations have the same value of $\underline{Y}$ within parentheses. In other words, the two observations have been made on the same fluid particle but at two different time instants: t and $t + \Delta t$. Thus, $\frac{\partial \phi^+(t,\underline{Y})}{\partial t}$ precisely represents the current rate (at time t) of change in the quantity ϕ following that fluid particle which was at location $\underline{Y}$ at the reference time t_{ref}.

Replacing $\phi^+(t, \underline{Y})$ by the velocity vector ($\underline{V}^+(t, \underline{Y})$ in (3.2) gives us the expression for the instantaneous acceleration of the fluid particle which was at location $\underline{Y}$ at the reference time t_{ref}:

$$\underline{a}^+(t, \underline{Y}) = \frac{\partial \underline{V}^+(t, \underline{Y})}{\partial t}. \tag{3.3}$$

Similarly, the partial derivatives $\frac{\partial \rho^+(t,\underline{Y})}{\partial t}$, $\frac{\partial p^+(t,\underline{Y})}{\partial t}$, and $\frac{\partial T^+(t,\underline{Y})}{\partial t}$ represent the instantaneous rates of change in density, pressure and temperature of that fluid particle which was located at $\underline{Y}$ at t_{ref}.

In the Lagrangian description, the current location of an independently chosen fluid particle (which had its location at $\underline{Y}$ at t_{ref}) is a dependent variable. This is represented as $\underline{X}^+(t, \underline{Y})$. Indeed, this quantity is related to the time integral of the velocity vector:

$$\underline{X}^+(t, \underline{Y}) = \underline{Y} + \int_{t_{ref}}^{t} \underline{V}^+(t^{'}, \underline{Y})\, dt^{'}, \tag{3.4}$$

where $t_{ref} \leq t^{'} \leq t$.

3.3 The Eulerian Description of Fluid Continuum

In the Eulerian description of a flow field, time (t) and an arbitrary position vector $\underline{X}$ (see Fig. 3.2) are treated as the set of independent variables. In turn, the velocity vector ($V(\underline{X}, t)$), density($\rho(\underline{X}, t)$), temperature ($T(\underline{X}, t)$), and pressure ($p(\underline{X}, t)$) of the fluid particle that happens to be located at $\underline{X}$ at the current time t are the dependent variables. It is conventional that, unlike Lagrangian dependent variables, the dependent variables of the Eulerian description do not have $+$ as a superscript. A further distinction is made by the order in which the independent variables appear as

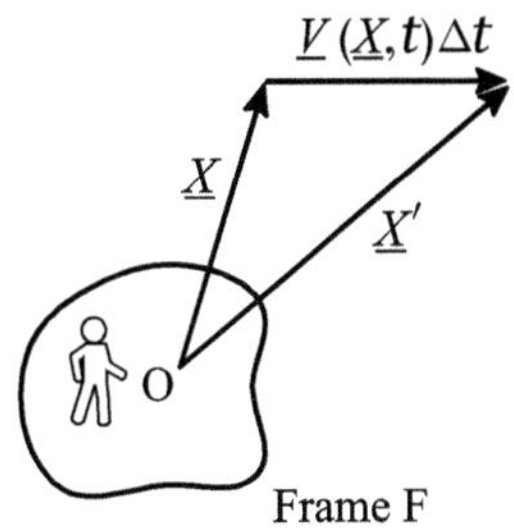

Fig. 3.2 The Eulerian description of fluid motion

arguments of the dependent variables. While the Lagrangian description had the time variable appearing first and the spatial variable ($\underline{Y}$) appearing second, the Eulerian description has the spatial variable ($\underline{X}$) appearing first and the time variable appearing second. Since $\underline{X}$ is a continuous spatial variable over our domain of interest, $V(\underline{X}, t)$, $\rho(\underline{X}, t)$, $T(\underline{X}, t)$, and $p(\underline{X}, t)$ are called the instantaneous Eulerian velocity, density, temperature, and pressure fields.

Finding the rate of change of a flow variable (velocity, density, temperature, and pressure) following a fluid particle is not as straightforward as it has been in the Lagrangian description, wherein merely taking a partial derivative of the flow variables was adequate (Eq. 3.2). In the Eulerian description, the partial derivative of a dependent variable, say $\phi(\underline{X}, t)$, with respect to time is:

$$\frac{\partial \phi(\underline{X}, t)}{\partial t} = \lim_{\Delta t \to 0} \frac{\phi(\underline{X}, t + \Delta t) - \phi(\underline{X}, t)}{\Delta t}. \tag{3.5}$$

Even though there are two measured values of ϕ in the numerator of rhs of (3.5), and indeed, they are at two different time instants (at t and $t + \Delta t$), they both are at the same spatial location $\underline{X}$. Since, in general, the fluid particle which is at $\underline{X}$ at time t is not stationary, when the next measurement of the quantity ϕ is made at the time instant at $t + \Delta t$, that fluid particle would move to another location, and a different fluid particle would be occupying the location $\underline{X}$. Thus the second measurement on the rhs of (3.5) is done on a different fluid particle. Thus, the partial derivative of $\phi(\underline{X}, t)$ does not represent the rate of change in ϕ associated with any particular fluid particle (neither the one which is at $\underline{X}$ at time t nor the one which would be at $\underline{X}$ at time $t + \Delta t$).

To derive the appropriate operator for the rate of change in quantity ϕ associated with a fluid particle which is at location $\underline{X}$ at time t, we refer to Fig. 3.2. The fluid particle, which is located at $\underline{X}$ at time t, moves with respect to the ground frame at the local instantaneous velocity vector $\underline{V}(\underline{X}, t)$. Thus, at time $t + \Delta t$ this fluid particle, being displaced by the local velocity vector $\underline{V}(\underline{X}, t)$, would move to the new location $\underline{X}'$, where

$$\underline{X}' \approx \underline{X} + V(\underline{X}, t)\Delta t. \tag{3.6}$$

Thus, the rate of change of ϕ associated with the same fluid particle which is at location $\underline{X}$ at time t must be mathematically represented as:

$$\lim_{\Delta t\to 0}\frac{\phi(\underline{X}',t+\Delta t)-\phi(\underline{X},t)}{\Delta t}=\lim_{\Delta t\to 0}\frac{\phi(\underline{X}+\underline{V}(\underline{X},t)\Delta t,t+\Delta t)-\phi(\underline{X},t)}{\Delta t}. \tag{3.7}$$

This expression can be further simplified by invoking the Taylor series approximation for $\phi(\underline{X}+\underline{V}(\underline{X},t)\Delta t,t+\Delta t)$. To understand this simplification process, let us employ a frame-fixed Cartesian coordinate system $Ox_1(\hat{e}_1)x_2(\hat{e}_2)x_3(\hat{e}_3)$to express various vectors:

$$\underline{X}=x_1\hat{e}_1+x_2\hat{e}_2+x_3\hat{e}_3,\ \text{and}$$
$$\underline{V}(\underline{X},t)=\underline{V}(x_1,x_2,x_3,t)=V_1\hat{e}_1+V_2\hat{e}_2+V_3\hat{e}_3, \tag{3.8}$$

where x_1, x_2, x_3 and V_1, V_2, V_3 represent the scalar components of the corresponding vectors. Accordingly, the quantity in (3.7) can be expressed as:

$$\lim_{\Delta t\to 0}\frac{\phi(x_1+V_1\Delta t,x_2+V_2\Delta t,x_3+V_3\Delta t,t+\Delta t)-\phi(x_1,x_2,x_3,t)}{\Delta t}. \tag{3.9}$$

Using the multivariate Taylor series approximation up to the first order, we make the following approximation:

$$\begin{aligned}\phi(x_1+V_1\Delta t,x_2+V_2\Delta t,x_3+V_3\Delta t,t+\Delta t)&\approx\phi(x_1,x_2,x_3,t)+\frac{\partial\phi}{\partial t}\Delta t\\&+\frac{\partial\phi}{\partial x_1}V_1\Delta t+\frac{\partial\phi}{\partial x_2}V_2\Delta t+\frac{\partial\phi}{\partial x_3}V_3\Delta t,\end{aligned} \tag{3.10}$$

where all partial derivatives appearing in (3.10) are at (x_1, x_2, x_3, t). Using (3.10) in (3.9) in the limit of $\Delta t\to 0$ leads to the following exact expression of the desired rate of change in $\phi(t, x_1, x_2, x_3)$:

$$\lim_{\Delta t\to 0}\frac{\phi(\underline{X}',t+\Delta t)-\phi(\underline{X},t)}{\Delta t}=\frac{\partial\phi}{\partial t}+V_1\frac{\partial\phi}{\partial x_1}+V_2\frac{\partial\phi}{\partial x_2}+V_3\frac{\partial\phi}{\partial x_3}. \tag{3.11}$$

The rhs of (3.11) involves not only the partial derivative of ϕ with respect to time but also the partial derivatives of the dependent variable with respect to three spatial coordinates x_1, x_2, x_3. Referring back to (1.58), (3.11) can be expressed in a form which is independent of the choice of coordinate system:

$$\begin{aligned}\lim_{\Delta t\to 0}\frac{\phi(\underline{X}',t+\Delta t)-\phi(\underline{X},t)}{\Delta t}&=\frac{\partial\phi(\underline{X},t)}{\partial t}+\left(\underline{V}\cdot\underline{\nabla}\right)\phi(\underline{X},t),\\&=\left(\frac{\partial}{\partial t}+\underline{V}\cdot\underline{\nabla}\right)\phi(\underline{X},t).\end{aligned} \tag{3.12}$$

In the study of fluid mechanics, the operator $\frac{\partial}{\partial t}+\left(\underline{V}\cdot\underline{\nabla}\right)$ is denoted by the symbol $\frac{D}{Dt}$, and is called the *material derivative* operator. It can act upon any dependent variable of the Eulerian description of a flow field. Here, the significance of the word "material" is that it represents the rate of change of the variable in context following the same fluid particle or material, which is at location $\underline{X}$ at time t.

The material derivative operator is applied on the Eulerian velocity field to arrive at the acceleration of the local fluid particle $\underline{a}(\underline{X}, t)$:

$$\underline{a}(\underline{X}, t) = \frac{D}{Dt}\underline{V}(\underline{X}, t). \tag{3.13}$$

The Lagrangian and the Eulerian descriptions of a continuum flow field offer their own individual advantages. The Lagrangian approach appears to be more intuitive in light of our fundamental training in classical particle mechanics, wherein, indeed, the governing equations of motion are directly written for independently chosen particles (rather than spatial locations). The particle's current location of interest is calculated as a dependent variable after determining its momentum. The mathematical process of finding the rate of change of properties associated with a chosen fluid particle too is mathematically simpler in the Lagrangian description (merely, the partial derivative with time) compared to the Eulerian description, which involves computation of both the time and the spatial derivatives (3.12). Still, the Eulerian description has been the more popular way of describing fluid motion. This preference is attributable to the fact that for most engineering problems, our focus is indeed to measure and understand the behavior of fluid at independently chosen locations/regions rather than the behavior of specific fluid particles. Further, the expression of the forces arising on a given fluid particle due to the interaction of the neighboring fluid particles (pressure forces and viscous forces, more on this in Chap. 5) involves spatial gradients of the dependent variables. The Eulerian description, with the spatial location being an independent variable, simplifies the algebraic expression and the manipulation of these gradients. In the rest of this book, we describe fluid motion using the Eulerian description only.

Chapter 4
Fluid Kinematics

The word *kinematics* means the study of the motion of a particle or a system of particles without making any reference to the forces that cause that motion. Thus, the phrase *fluid kinematics* means the study of the motion of fluid particles or that of a system of fluid particles without making any reference to the forces that cause that motion.

In the Eulerian description of a flow field, we express the velocity field over the domain of interest as a function of space ($\underline{X}$) and time (t): $\underline{V}(\underline{X}, t)$. A spatially varying Eulerian velocity field is called a *non-uniform* velocity field, whereas a time-dependent Eulerian velocity field is called an *unsteady* velocity field. In contrast, the Eulerian velocity field is described as *steady* if there is no time dependence. Similarly, if an Eulerian velocity field has no dependence on space, it is called a *uniform* Eulerian velocity field.

Consider an Eulerian velocity field that is expressed using a coordinate system $Ox_1(\hat{e}_1)x_2(\hat{e}_2)x_3(\hat{e}_3)$ (Fig. 1.1). If one particular scalar component of the velocity field is zero at all locations, then such a velocity field is called a *two-component* or *2C* velocity field. Similarly, a *1-component* or *1C* velocity field may exist with two particular scalar components being identically zero at all locations.

If an Eulerian velocity vector field depends only on two coordinates of the working coordinate system, it is called a two-dimensional or 2D velocity field. Similarly, a one-dimensional (or 1D) velocity field is one in which the velocity field depends only on one coordinate. In general, a velocity field may depend on all three coordinates; thus, a velocity field is three-dimensional or 3D. The following is an example of a velocity field which is two-dimensional (2D) but has three components (3C).

$$\underline{V} = (\alpha x_1 x_2 \hat{e}_1 + \beta x_2 \hat{e}_2 + \gamma x_1^2 \hat{e}_3). \tag{4.1}$$

The following is an example of a velocity field which is three-dimensional (3D) but has two components (2C)

© The Author(s), under exclusive license to Springer Nature Switzerland AG 2026

S. S. Sinha, *Fundamentals of Fluid Mechanics*,
https://doi.org/10.1007/978-3-032-04210-1_4

$$\underline{V} = (\alpha x_1 x_2 \hat{e}_1 + \beta x_2 x_3 \hat{e}_2), \tag{4.2}$$

where α, β and γ are constants.

If an explicit mathematical function for $\underline{V}(\underline{X}, t)$ is available, one can find the instantaneous velocity vector of each fluid particle present in the domain of interest. A simple example of such a velocity field is

$$\underline{V}(x_1, x_2, x_3, t) = t(\alpha x_1 \hat{e}_1 + \beta x_2 \hat{e}_2 + \gamma x_3 \hat{e}_3), \tag{4.3}$$

where α, β and γ are constants. With such a known mathematical expression of the velocity field, the acceleration field can be readily derived using the material derivative operator described in Chap. 3 (Eq. 3.13)

$$\begin{aligned} \underline{a}(\underline{X}, t) &= \left[\frac{\partial}{\partial t} + V_i \frac{\partial}{\partial x_i}\right] \underline{V}, \\ &= \left[\frac{\partial}{\partial t}\right] \underline{V} + \left[V_1 \frac{\partial}{\partial x_1} + V_2 \frac{\partial}{\partial x_2} + V_3 \frac{\partial}{\partial x_3}\right] \underline{V}, \\ &= \alpha x_1 (t^2 \alpha + 1)\hat{e}_1 + \beta x_2 (t^2 \beta + 1)\hat{e}_2 \\ &\quad + \gamma x_3 (t^2 \gamma + 1)\hat{e}_3. \end{aligned} \tag{4.4}$$

Although the description of an Eulerian velocity field may be complete by its mathematical expression, such as (4.3), it may not always be easy to visualize the flow field by merely looking at its mathematical expression. Thus, there exist some graphical techniques which can help us visualize the flow field with relative ease. We study three such visualization techniques in this chapter: *streamlines*, *streaklines*, and *pathlines*.

4.1 Streamlines, Streaklines, and Pathlines

A *streamline* is a curve in the fluid domain such that the local tangent at every point on this curve points in the direction of the local instantaneous velocity vector. Figure 4.1 shows some representative streamlines in a flow field over a curved solid surface. It is conventional to annotate the streamlines with arrowheads to denote the direction of the flow. In Fig. 4.2, we focus on the fluid particle A, which currently has spatial coordinates (x_1, x_2, x_3). This fluid particle currently lies on the shown streamline, $P'P''$, which, in general, is a curve in the three-dimensional space. If $\Delta \underline{l}$ represents a vector of infinitesimal length along the streamline (where the particle A is currently located), then

$$\lim_{\Delta \underline{l} \to 0} \frac{\Delta l_1}{\Delta l_2} = \frac{V_1}{V_2}, \lim_{\Delta \underline{l} \to 0} \frac{\Delta l_2}{\Delta l_3} = \frac{V_2}{V_3}, \lim_{\Delta \underline{l} \to 0} \frac{\Delta l_3}{\Delta l_1} = \frac{V_3}{V_1}, \tag{4.5}$$

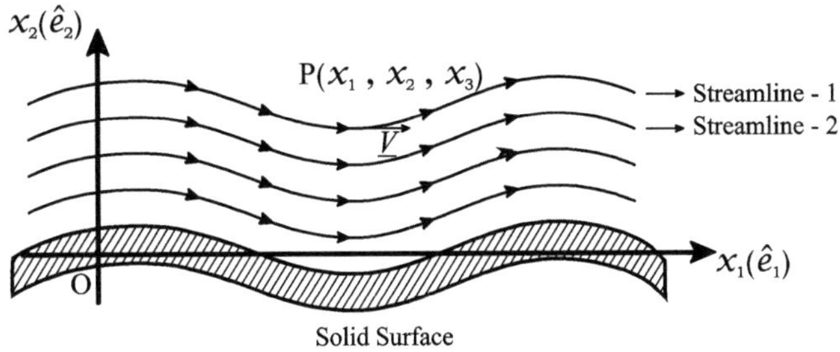

Fig. 4.1 Streamlines in a flow domain

Fig. 4.2 An infinitesimal displacement vector on the streamline S_1 at location A with coordinates (x_1, x_2, x_3)

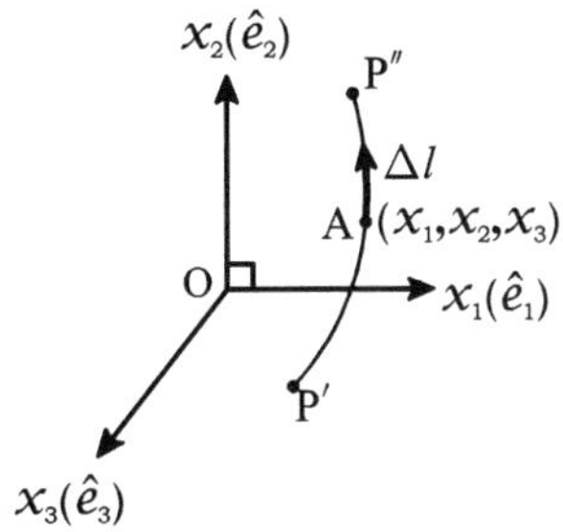

where $\Delta \underline{l} = \Delta l_1 \hat{e}_1 + \Delta l_2 \hat{e}_2 + \Delta l_3 \hat{e}_3$ and Δl_1, Δl_2, Δl_3 are the scalar components of the vector. The symbols V_1, V_2, and V_3 denote the scalar components of the velocity vector at location A.

At a given instant, a streamline in a flow field can be generated by first selecting a location S_{ref}, which can be called the *seeding location* for that streamline. The seeding location must have a well-defined non-zero velocity vector. Tracing the direction of the known instantaneous local velocity vector, an infinitesimal increment $\Delta \underline{l}$ can then be made to extend the streamline to another discrete location in the flow field. Multiple repetitions of this step can then be used to extend the streamline curve across the domain. In the limit of each segment tending toward zero, a smooth curve can be obtained. This smooth curve is the unique streamline that exists, passing through the location S_{ref} at the given time instant.

In an unsteady flow field, different streamline curves will exist at different time instants, albeit passing through the same location S_{ref}. However, in a steady velocity field, a unique, time-independent streamline passes through S_{ref}. No streamline passes through a location where the flow is stagnant (the instantaneous local velocity vector direction itself is not defined). Further, no two streamlines can ever intersect because having two different directions of the velocity vector at the same location is impossible.

Based on the definition of a streamline, we define *stream-surfaces* and *stream-tubes*. A stream surface is an imaginary surface made up of contiguous streamlines. Since the local velocity vector is always tangent to a streamline, a stream surface is impervious to fluid flow (Fig. 4.3). No fluid particle can ever pass through a stream

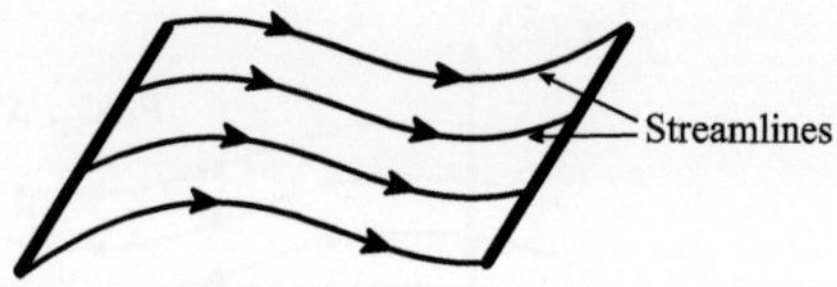

Fig. 4.3 A stream surface

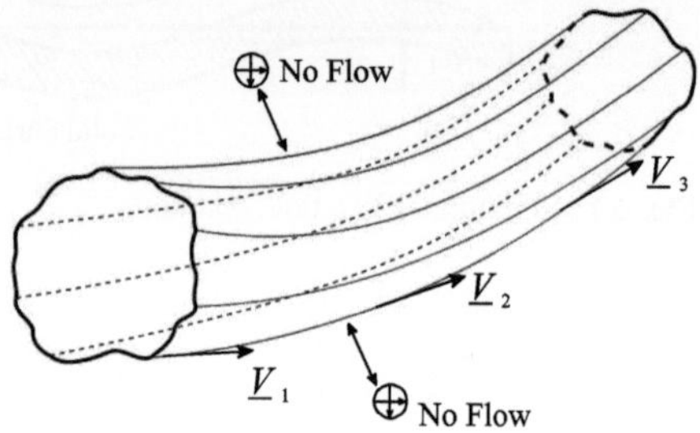

Fig. 4.4 A stream-tube

surface. On the other hand, a stream-tube is a closed stream surface such that it forms an imaginary conduit in the flow field (Fig. 4.4). Again, following the definition of a streamline, no flow can pass through the walls of a stream-tube. Stream-surfaces and stream-tubes not only help in flow visualization, but they prove useful in performing some insightful theoretical analyses of flow fields (more on this in Chap. 9).

A *streakline* is the instantaneous (at the current time instant) locus of all those fluid particles which passed through a fixed location (say L_{ref}) at various time instants in the past. If we place a dye releaser in a flow field at a spatial location (L_{ref}), then all those fluid particles which pass through L_{ref} at different times are expected to be visually distinguishable at later times. At an arbitrary time instant, the locus of all those dyed particles would appear as a curve, which would be the streakline associated with that location L_{ref}.

A *path-line* is the trajectory of a chosen fluid particle. This curve connects all those spatial locations where a chosen fluid particle has been since the time the flow field has been observed.

In a flow with a steady velocity field, the streakline associated with a reference location (L_{ref}) is also a streamline passing through the same location. Further, the same curve is also the path-line of all the fluid particles that are located on it.

4.2 A Fluid Element

The motion of a body can be completely described by tracking the motion of each of its constituent particles. Since inter-particle distances never change for a rigid body, we can define a unique vector representing the angular motion of the entire rigid body: the angular velocity vector. The angular velocity vector is defined as the rate (with respect to time) at which any arbitrarily chosen line segment of the rigid body changes its orientation with respect to the reference frame in context. With the

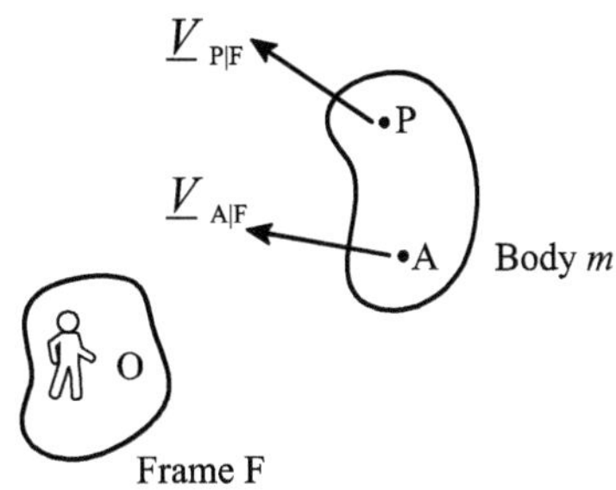

Fig. 4.5 Velocity of a typical particle P which is fixed to the rigid body m

known angular velocity vector of the rigid body and the known velocity of any one point (say point A) of a rigid body, the velocity of every other point (say P) can be ascertained (see Fig. 4.5):

$$\underline{V}_{P|F} = \underline{V}_{A|F} + \underline{\omega}_{body|F} \times \underline{r}_{PA}, \tag{4.6}$$

where F is the reference frame in context, and $\underline{\omega}_{body|F}$ represents the angular velocity of the rigid body with respect to frame F.

However, for a non-rigid body (or a deformable body), like a chunk of fluid mass, this approach (4.6) is not applicable. For such deformable bodies, no unique angular velocity vector exists. In general, different line segments of a non-rigid body may change their orientation at different rates in time. This difference between a deformable body and a rigid body arises because some or all inter-particle distances in a non-rigid body may change with time. Thus, to track the motion of such a non-rigid piece of mass, we have to keep track of the motion of all constituent fluid particles simultaneously using their individual velocity vectors.

Even though tracking individual fluid particles can completely describe the flow field, examining how a piece of fluid mass deforms in the flow domain is still insightful. This deformation is quantified in terms of the rates at which various inter-particle distances change with respect to time within a small piece of fluid mass (comprised of several fluid particles) and the rates at which the orientations of various line segments connecting these particles change within the same fluid mass with respect to time. Such measures of deformation can provide more insights into the fluid motion, and we will learn in Chap. 5 that some of these measures of deformation are related to the development of internal forces in a fluid continuum. Our study will eventually require estimating these internal forces to derive the governing equation of motion.

To quantify the deformation process of a fluid medium, we now introduce the idea of a *fluid element*. A fluid element is a small but finite-sized mass comprised of numerous fluid particles. We identify a fluid element at a reference time (t), then examine how it subsequently deforms over an infinitesimal duration Δt. Figure 4.6 shows a fluid element identified at time t. Our working coordinate system is a frame-fixed Cartesian coordinate system $Ox_1(\hat{e}_1)x_2(\hat{e}_2)x_3(\hat{e}_3)$.

At the current time t this fluid element has a cuboidal shape with its vertices being ABCDEFGH and its edge lengths are small but finite: Δx_1, Δx_2 and Δx_3 along the

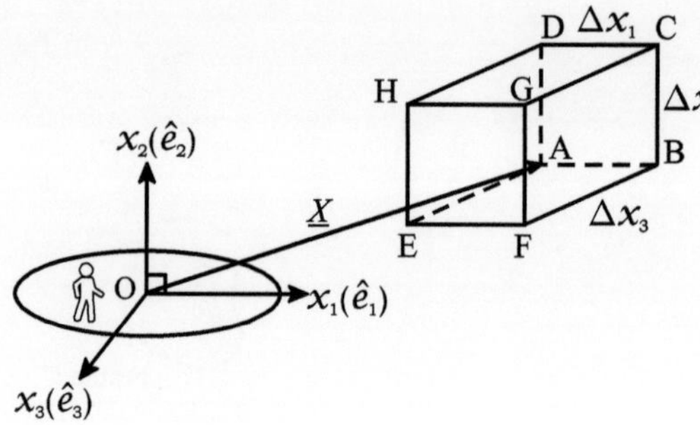

Fig. 4.6 A fluid element ABCDEFGH

directions $\hat{e}_1$, $\hat{e}_2$, and $\hat{e}_3$, respectively. At the current time instant, the coordinates of the six vertices are:

$$\begin{aligned}
&\text{Vertex A} : (x_1, x_2, x_3),\\
&\text{Vertex B} : (x_1 + \Delta x_1, x_2, x_3),\\
&\text{Vertex C} : (x_1 + \Delta x_1, x_2 + \Delta x_2, x_3),\\
&\text{Vertex D} : (x_1, x_2 + \Delta x_2, x_3),\\
&\text{Vertex E} : (x_1, x_2, x_3 + \Delta x_3),\\
&\text{Vertex F} : (x_1 + \Delta x_1, x_2, x_3 + \Delta x_3),\\
&\text{Vertex G} : (x_1 + \Delta x_1, x_2 + \Delta x_2, x_3 + \Delta x_3),\\
&\text{Vertex H} : (x_1, x_2 + \Delta x_1, x_3 + \Delta x_3).
\end{aligned} \tag{4.7}$$

Figure 4.7 shows the same fluid element at an infinitesimal time (Δt) later (at $t + \Delta t$). In general, the six vertices have their coordinates changed compared to what they had at t. The displacement of each vertex is governed by the product of the instantaneous local velocity vector at time t and the time increment Δt.

4.3 A Deforming Fluid Element in a 2C, 2D Velocity Field

For algebraic simplicity at this point, let us assume that the background Eulerian velocity field is 2C ($V_3 = 0$) everywhere in the flow field and that it is 2D such that V_1 and V_2 do not depend on x_3. In such a flow field, the face ABCD of the fluid element always remains in the plane $x_3 = 0$. Similarly, the face EFGH always remains in the plane $x_3 = \Delta x_3$. With time, the changes occurring to face ABCD of the fluid element would be identical to the changes that happen to face EFGH. Thus, for such a flow field, focusing on the face ABCD alone is adequate. In Fig. 4.8a, b, we present the states of face ABCD at t and $t + \Delta t$, respectively.

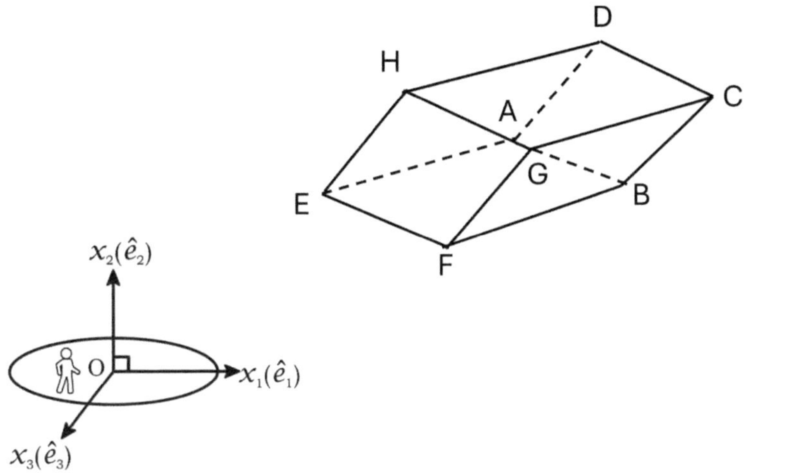

Fig. 4.7 The shape of the fluid element ABCDEFGH at $t + \Delta t$. The shape of the same fluid element was a cuboid at time t (Fig. 4.6)

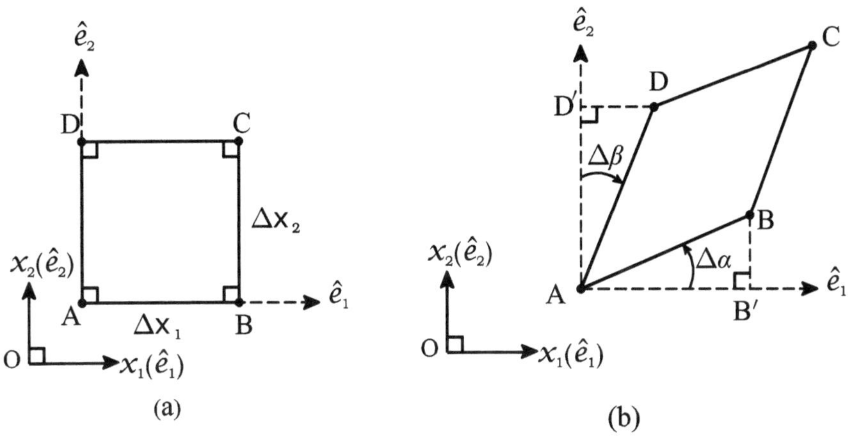

Fig. 4.8 The shape of the face ABCD of the fluid element ABCDEFGH at time $t + \Delta t$ (sub-figure b) in a 2C, 2D velocity field. The shape of the face was a rectangle at time t (sub-figure a)

In Fig. 4.8 two angles (measured in radians) are marked: $\Delta\alpha$ and $\Delta\beta$. These represent changes in the orientation of the edges AB and AD at time $t + \Delta t$ relative to the respective orientations of these edges at t. While $\Delta\alpha$ is measured positive counter-clockwise, $\Delta\beta$ is measured positive clockwise.

With reference to Fig. 4.8, we define the *fractional rate of extension of edge AB* as:

$$\lim_{\Delta t \to 0} \frac{(AB)_{t+\Delta t} - (AB)_t}{(AB)_t \Delta t}, \tag{4.8}$$

where $(AB)_t$ and $(AB)_{t+\Delta t}$ denote the lengths of edge AB at t and $t + \Delta t$, respectively. Similarly, the *fractional rate of extension of the edge AD* is expressed as

$$\lim_{\Delta t \to 0} \frac{(AD)_{t+\Delta t} - (AD)_t}{(AD)_t \Delta t}. \tag{4.9}$$

The *rate of fractional change in volume* of the fluid element ABCDEFGH is identified as:

$$\lim_{\Delta t \to 0} \frac{1}{\Delta t} \frac{vol_{t+\Delta t} - vol_t}{vol_t}, \tag{4.10}$$

where vol_t and $vol_{t+\Delta t}$ represent the volume of the fluid element at t and $t + \Delta t$, respectively. This quantity (4.10) is also referred to as the dilatation rate of the fluid element.

The rate at which the edges AB and AD (which are oriented along the unit vector $\hat{e}_1$ and $\hat{e}_2$, respectively) tend to align toward each other equals:

$$\lim_{\Delta t \to 0} \frac{\Delta \alpha + \Delta \beta}{\Delta t}. \tag{4.11}$$

Equations (4.8)–(4.11) are indeed consequences of the non-rigid (or the deformable) nature of the fluid element. If the fluid element ABCDEFGH was a rigid body, all these quantities defined in (4.8)–(4.11) would be identically zero. Since a fluid element is generally a non-rigid body, different line segments on the body may have different rates of change in their orientations. Specifically, in the context of Fig. 4.8, this means, in general, $\Delta \alpha \neq -\Delta \beta$. Thus, we cannot describe the rate of change in orientation of the face ABCD in terms of any uniquely definable angular velocity vector. Nonetheless, in fluid kinematics, we still define *the averaged angular velocity vector*, $\underline{\Omega}$ of a fluid element. Based on Fig. 4.8, we define the component of the averaged angular velocity vector along the $\hat{e}_3$ unit vector as

$$\Omega_3 = \lim_{\Delta t \to 0} \frac{\Delta \alpha - \Delta \beta}{2 \Delta t}. \tag{4.12}$$

The factor 2 introduced in (4.12) is to calculate the *averaged* rate of change in orientation of two line segments AB and CD of the body. This definition is mathematically different from the angular velocity component of a rigid body, which would have been

$$\Omega_3^{rigidbody} = \lim_{\Delta t \to 0} \frac{\Delta \alpha}{\Delta t} = - \lim_{\Delta t \to 0} \frac{\Delta \beta}{\Delta t}. \tag{4.13}$$

Despite the difference, (4.12) is still deemed useful in gauging an *averaged* rate of change in the orientation of the fluid element. If the fluid element ABCDEFGH instantaneously behaves like a rigid body (may happen in some flow fields), $\Delta \beta = -\Delta \alpha$, and the definition of the averaged angular velocity component in (4.12) becomes exactly equal to the third component of the unique angular velocity vector ($\Omega_3^{rigidbody}$) of the rigid fluid element (4.13).

We now wish to find the relationships between the geometrical quantities defined in (4.8)–(4.12) and the background velocity field in the flow domain. This requires determination of the coordinates of the vertices A, B, C, and D (Fig. 4.8) at t as well as those at $t + \Delta t$. The coordinates of vertices ABCD are already known based on the identified shape and the size of the fluid element at time t (4.7). To find the coordinates of the vertices at $t + \Delta t$, we use the approximation that for a small increment Δt, the corresponding displacement vector of a vertex can be approximated as the product of Δt with the local velocity vector existing at the location of that vertex at time t. This approximation becomes exact in the limit of Δt approaching zero.

Since the fluid element is small (Δx_1, Δx_2, and Δx_3 are all small), the velocity components at vertices B and D can be expressed using a truncated (up to first order) Taylor series expansion with vertex A being as the base location.

Vertex A:

$$\underline{V}(x_1, x_2, x_3, t) = V_1\hat{e}_1 + V_2\hat{e}_2, \tag{4.14}$$

Vertex B:

$$\begin{aligned}\underline{V}(x_1 + \Delta x_1, x_2, x_3, t) &= \left(V_1 + \frac{\partial V_1}{\partial x_1}\Delta x_1 + \text{H.O.T.}\right)\hat{e}_1 \\ &\quad + \left(V_2 + \frac{\partial V_2}{\partial x_1}\Delta x_1 + \text{H.O.T.}\right)\hat{e}_2 \\ &\approx \left(V_1 + \frac{\partial V_1}{\partial x_1}\Delta x_1\right)\hat{e}_1 + \left(V_2 + \frac{\partial V_2}{\partial x_1}\Delta x_1\right)\hat{e}_2,\end{aligned} \tag{4.15}$$

Vertex D:

$$\begin{aligned}&\underline{V}(x_1, x_2 + \Delta x_2, x_3, t) = \\ &\quad \left(V_1 + \frac{\partial V_1}{\partial x_2}\Delta x_2 + \text{H.O.T.}\right)\hat{e}_1 + \left(V_2 + \frac{\partial V_2}{\partial x_2}\Delta x_2 + \text{H.O.T.}\right)\hat{e}_2 \\ &\quad \approx \left(V_1 + \frac{\partial V_1}{\partial x_2}\Delta x_2\right)\hat{e}_1 + \left(V_2 + \frac{\partial V_2}{\partial x_2}\Delta x_2\right)\hat{e}_2,\end{aligned} \tag{4.16}$$

where the symbols V_1, V_2, V_3, and all their partial derivatives are computed at the base location (vertex A) with coordinates (x_1, x_2, x_3) at time t, and H.O.T. means higher order terms of the relevant Taylor series expansion.

Using the coordinates of vertices A, B and D at time t (as listed in 4.7) and taking products of the respective velocity vectors (4.14)–(4.16) with the small-time increment Δt, the coordinates of the vertices A, B, and D at $t + \Delta t$ are computed as

$$
\begin{aligned}
&\text{Vertex A}\\
&x_1 + V_1 \Delta t,\ x_2 + V_2 \Delta t,\ x_3,\\
&\text{Vertex B}\\
&x_1 + \Delta x_1 + \left(V_1 + \frac{\partial V_1}{\partial x_1} \Delta x_1 \right) \Delta t,\ x_2 + \left(V_2 + \frac{\partial V_2}{\partial x_1} \Delta x_1 \right) \Delta t, x_3,\\
&\text{Vertex D}\\
&x_1 + \left(V_1 + \frac{\partial V_1}{\partial x_2} \Delta x_2 \right) \Delta t,\ x_2 + \Delta x_2 + \left(V_2 + \frac{\partial V_2}{\partial x_2} \Delta x_2 \right) \Delta t, x_3.
\end{aligned}
\tag{4.17}
$$

Using (4.17) and the known coordinates of A, B, and D at time t (4.7), we can now find the expressions of rates of various geometric changes of the fluid element (4.8)–(4.12) in terms of the background velocity field or its spatial derivatives.

With reference to Fig. 4.8, the rate of fractional change in the length of AB (4.8) is first approximated as

$$
\lim_{\Delta t \to 0} \frac{1}{\Delta t} \frac{(AB)_{t+\Delta t} - (AB)_t}{(AB)_t} = \lim_{\Delta t \to 0} \frac{1}{\Delta t} \frac{(AB')_{t+\Delta t} - (AB)_t}{(AB)_t}. \tag{4.18}
$$

Here $AB^{'}$ is the projection of AB along the $\hat{e}_1$ direction (Fig. 4.8), and we are approximating $(AB)_{t+\Delta t} \approx (AB^{'})_{t+\Delta t}$. This approximation is reasonable, because for an infinitesimal Δt, $\Delta \alpha$, is small too and $(AB^{'})_{t+\Delta t} = (AB)_{t+\Delta t} cos \Delta \alpha \approx (AB)_{t+\Delta t}$. Using the coordinates listed in (4.17) leads to

$$
\begin{aligned}
\lim_{\Delta t \to 0} \frac{1}{\Delta t} \frac{(AB)_{t+\Delta t} - (AB)_t}{(AB)_t} &= \lim_{\Delta t \to 0} \frac{1}{\Delta t} \frac{(AB')_{t+\Delta t} - (AB)_t}{(AB)_t},\\
&= \lim_{\Delta t \to 0} \frac{1}{\Delta t} \frac{(\Delta x_1 + \frac{\partial V_1}{\partial x_1} \Delta x_1 \Delta t) - (\Delta x_1)}{\Delta x_1},\\
&= \frac{\partial V_1}{\partial x_1}.
\end{aligned}
\tag{4.19}
$$

Similarly, leveraging the fact that $\Delta \beta$, too, is a small angle, the rate of fractional change in the length of AD is expressed as:

$$
\begin{aligned}
\lim_{\Delta t \to 0} \frac{1}{\Delta t} \frac{(AD)_{t+\Delta t} - (AD)_t}{(AD)_t} &= \lim_{\Delta t \to 0} \frac{1}{\Delta t} \frac{(AD')_{t+\Delta t} - (AD)_t}{(AD)_t},\\
&= \lim_{\Delta t \to 0} \frac{1}{\Delta t} \frac{(\Delta x_2 + \frac{\partial V_2}{\partial x_2} \Delta x_2 \Delta t) - (\Delta x_2)}{\Delta x_2},\\
&= \frac{\partial V_2}{\partial x_2},
\end{aligned}
\tag{4.20}
$$

where $AD^{'}$ is the projection of AD along the $\hat{e}_2$ direction.

To determine the volume of the fluid ABCDEFGH at $t + \Delta t$, again we use the small-angle approximations for $\Delta\alpha$ and $\Delta\beta$ to arrive at:

$$
\begin{aligned}
Vol_{t+\Delta t} &\approx (AB)_{t+\Delta t} \times (AD)_{t+\Delta t} \times \Delta x_3,\\
&\approx (AB')_{t+\Delta t} \times (AD')_{t+\Delta t} \times \Delta x_3.
\end{aligned} \tag{4.21}
$$

Thus, the dilatation rate of the fluid element can be expressed as:

$$
\begin{aligned}
\lim_{\Delta t \to 0} \frac{1}{\Delta t} \frac{vol_{t+\Delta t} - vol_t}{vol_t} &=\\
&\lim_{\Delta t \to 0} \frac{1}{\Delta t} \frac{(AB)_{t+\Delta t}(AD)_{t+\Delta t}\Delta x_3 - (AB)_t(AD)_t\Delta x_3}{(AB)_t(AD)_t\Delta x_3},\\
&= \lim_{\Delta t \to 0} \frac{1}{\Delta t} \frac{(AB')_{t+\Delta t}(AD')_{t+\Delta t}\Delta x_3 - (AB)_t(AD)_t\Delta x_3}{(AB)_t(AD)_t\Delta x_3},\\
&= \lim_{\Delta t \to 0} \frac{1}{\Delta t \Delta x_1 \Delta x_2 \Delta x_3} \left\{ \Delta x_1 \Delta x_2 \Delta x_3 + \left(\frac{\partial V_1}{\partial x_1} + \frac{\partial V_2}{\partial x_2} \right.\right. \qquad (4.22)\\
&\qquad \left.\left. + \frac{\partial V_1}{\partial x_1}\frac{\partial V_2}{\partial x_2}\Delta t \right) \Delta x_1 \Delta x_2 \Delta x_3 \Delta t - \Delta x_1 \Delta x_2 \Delta x_3 \right\},\\
&= \frac{\partial V_1}{\partial x_1} + \frac{\partial V_2}{\partial x_2}. \qquad (4.23)
\end{aligned}
$$

Since $\Delta\alpha$ and $\Delta\beta$ are small angles, these angles, themselves, can be expressed in terms of the coordinates of various vertices listed in (refer to Eq. 4.17 and Fig. 4.8):

$$
\begin{aligned}
\Delta\alpha \approx tan\Delta\alpha &= \frac{BB'}{AB'},\\
&\approx \frac{[x_2 + (V_2 + \frac{\partial V_2}{\partial x_1}\Delta x_1)\Delta t] - [x_2 + V_2\Delta t]}{[x_1 + \Delta x_1 + (V_1 + \frac{\partial V_1}{\partial x_1}\Delta x_1)\Delta t] - [x_1 + V_1\Delta t]},\\
&\approx \frac{\frac{\partial V_2}{\partial x_1}\Delta t}{1 + \frac{\partial V_1}{\partial x_1}\Delta t},
\end{aligned} \tag{4.24}
$$

$$
\begin{aligned}
\Delta\beta \approx tan\Delta\beta &= \frac{DD'}{AD'},\\
&\approx \frac{[x_1 + (V_1 + \frac{\partial V_1}{\partial x_2}\Delta x_2)\Delta t] - [x_1 + V_1\Delta t]}{[x_2 + \Delta x_2 + (V_2 + \frac{\partial V_2}{\partial x_2}\Delta x_2)\Delta t] - [x_2 + V_2\Delta t]},\\
&\approx \frac{\frac{\partial V_1}{\partial x_2}\Delta t}{1 + \frac{\partial V_2}{\partial x_2}\Delta t}.
\end{aligned} \tag{4.25}
$$

Using (4.11) along with (4.24) and (4.25), the rate at which edges AB and AD tend to align with each other is expressed as:

$$\lim_{\Delta t \to 0} \frac{\Delta\alpha + \Delta\beta}{\Delta t} = \frac{\partial V_2}{\partial x_1} + \frac{\partial V_1}{\partial x_2}. \tag{4.26}$$

Finally, using (4.24) and (4.25) in (4.12), the third component of the averaged angular velocity vector, Ω_3, is expressed as:

$$\lim_{\Delta t \to 0} \frac{\Delta\alpha - \Delta\beta}{2\Delta t} = \frac{1}{2}\left(\frac{\partial V_2}{\partial x_1} - \frac{\partial V_1}{\partial x_2}\right). \tag{4.27}$$

4.4 A Deforming Fluid Element in a 3C, 3D Velocity Field

In Sect. 4.3, we restricted ourselves to a 2C, 2D flow field for algebraic simplicity and related the rates of various geometric changes happening to a fluid element to the velocity field and its spatial derivatives. The insights gained from a 2C, 2D velocity field can now be extended to a general 3C, 3D velocity field. We go back to the fluid element of Fig. 4.7 at $t + \Delta t$. Figure 4.6 shows the corresponding undeformed state of the fluid element. Since $V_3 \neq 0$, the constituent fluid particles of the fluid element would also be displaced in the $\hat{e}_3$ direction. In general, the vertices A, B, C, and D would no longer be lying in a plane parallel to the $x_1(\hat{e}_1) - x_2(\hat{e}_2)$ plane. Similarly, the vertices, E, F, G, and H, would no more be confined to a plane parallel to the $x_1(\hat{e}_1) - x_2(\hat{e}_2)$ plane. In Fig. 4.9, we have shown projections of edges AB, AD, and AE of the fluid element on three orthogonal planes at time $t + \Delta t$. These segments have not been shown to avoid clutter in the figure. The projections of segments AB and AD on the $x_1(\hat{e}_1) - x_2(\hat{e}_2)$ plane are AB' and AD', respectively. Similarly, the projections of segments AD and AE on the $x_2(\hat{e}_2) - x_3(\hat{e}_3)$ plane are AD'' and AE'', respectively. The projections of segments AE and AB on the $x_3(\hat{e}_3) - x_1(\hat{e}_1)$ plane are AE''' and AB''', respectively. Further, on these figures, we have shown various small angles that these projected line segments make with the three axes ($x_1(\hat{e}_1)$, $x_2(\hat{e}_2)$, and $x_3(\hat{e}_3)$). Using Fig. 4.9 as a reference and the expressions obtained earlier in (4.19)–(4.27), we can derive the various rates of geometric changes associated with the fluid element.

1. The rate of fractional change in the length AB is derived as:

$$\lim_{\Delta t \to 0} \frac{1}{\Delta t} \frac{(AB)_{t+\Delta t} - (AB)_t}{(AB)_t} = \frac{\partial V_1}{\partial x_1}. \tag{4.28}$$

2. The rate of fractional change in the length AD is derived as:

$$\lim_{\Delta t \to 0} \frac{1}{\Delta t} \frac{(AD)_{t+\Delta t} - (AD)_t}{(AD)_t} = \frac{\partial V_2}{\partial x_2}. \tag{4.29}$$

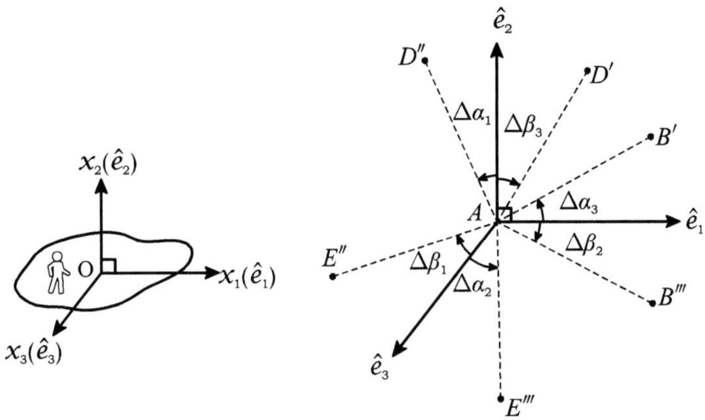

Fig. 4.9 Different projections of the edges AB, AD, and AE of the fluid element ABCDEFGH on the three planes of the working coordinate system at time $t + \Delta t$. Relevant small angles and their directions are also marked on the figure. The same fluid element was a cuboid at time t (Fig. 4.6)

3. The rate of fractional change in the length AE is derived as:

$$\lim_{\Delta t \to 0} \frac{1}{\Delta t} \frac{(AE)_{t+\Delta t} - (AE)_t}{(AE)_t} = \frac{\partial V_3}{\partial x_3}. \tag{4.30}$$

4. The rate of fractional change in the volume (dilatation rate) of the fluid element ABCDEFGH is derived as:

$$\lim_{\Delta t \to 0} \frac{1}{\Delta t} \frac{vol_{t+\Delta t} - vol_t}{vol_t} = \frac{\partial V_1}{\partial x_1} + \frac{\partial V_2}{\partial x_2} + \frac{\partial V_3}{\partial x_3}. \tag{4.31}$$

5. The rate at which the projections of AB and AD on the $x_1(\hat{e}_1) - x_2(\hat{e}_2)$ plane tend to align with each other is derived as:

$$\lim_{\Delta t \to 0} \frac{\Delta\alpha_3 + \Delta\beta_3}{\Delta t} = \frac{\partial V_2}{\partial x_1} + \frac{\partial V_1}{\partial x_2}. \tag{4.32}$$

6. The rate at which the projections of AD and AE on the $x_2(\hat{e}_2) - x_3(\hat{e}_3)$ plane tend to align with each other is derived as:

$$\lim_{\Delta t \to 0} \frac{\Delta\alpha_1 + \Delta\beta_1}{\Delta t} = \frac{\partial V_3}{\partial x_2} + \frac{\partial V_2}{\partial x_3}. \tag{4.33}$$

7. The rate at which the projections of AE and AB on the $x_3(\hat{e}_3) - x_1(\hat{e}_1)$ plane tend to align with each other is derived as

$$\lim_{\Delta t \to 0} \frac{\Delta\alpha_2 + \Delta\beta_2}{\Delta t} = \frac{\partial V_1}{\partial x_3} + \frac{\partial V_3}{\partial x_1}. \tag{4.34}$$

8. The component of the averaged angular velocity vector along the unit vector $\hat{e}_3$ is derived as

$$\Omega_3 = \lim_{\Delta t \to 0} \frac{\Delta\alpha_3 - \Delta\beta_3}{2\Delta t} = \frac{1}{2}\left(\frac{\partial V_2}{\partial x_1} - \frac{\partial V_1}{\partial x_2}\right). \tag{4.35}$$

9. The component of the averaged angular velocity vector along the unit vector $\hat{e}_1$ is derived as

$$\Omega_1 = \lim_{\Delta t \to 0} \frac{\Delta\alpha_1 - \Delta\beta_1}{2\Delta t} = \frac{1}{2}\left(\frac{\partial V_3}{\partial x_2} - \frac{\partial V_2}{\partial x_3}\right). \tag{4.36}$$

10. The component of the averaged angular velocity vector along the unit vector $\hat{e}_2$ is derived as

$$\Omega_2 = \lim_{\Delta t \to 0} \frac{\Delta\alpha_2 - \Delta\beta_2}{2\Delta t} = \frac{1}{2}\left(\frac{\partial V_1}{\partial x_3} - \frac{\partial V_3}{\partial x_1}\right). \tag{4.37}$$

With its three scalar components (Ω_1, Ω_2, and Ω_3) the averaged angular velocity vector itself is expressed as: $\underline{\Omega} = \Omega_1\hat{e}_1 + \Omega_2\hat{e}_2 + \Omega_3\hat{e}_3$. It can be easily verified that this quantity is related to the curl of the velocity field as:

$$\underline{\Omega} = \frac{1}{2}\left(\underline{\nabla} \times \underline{V}\right). \tag{4.38}$$

In fluid mechanics, the quantity $\underline{\nabla} \times \underline{V}$ is also called *the vorticity vector.*

Further, the Cartesian components of the angular velocity vector are also related to the Cartesian components of the so-called *rotation-rate tensor* ($\underline{\mathrm{R}}$). $\underline{R}$ is a second-order tensor and is the antisymmetric part of the velocity gradient tensor($\underline{\nabla V}$):

$$\underline{R} = \frac{(\nabla V) - (\nabla V)^T}{2}. \tag{4.39}$$

and

$$\underline{R} = \Omega_3\hat{e}_1\hat{e}_2 - \Omega_2\hat{e}_1\hat{e}_3 - \Omega_3\hat{e}_2\hat{e}_1 + \Omega_1\hat{e}_2\hat{e}_3 + \Omega_2\hat{e}_3\hat{e}_1 - \Omega_1\hat{e}_3\hat{e}_2. \tag{4.40}$$

On the other hand, the symmetric part of the velocity gradient tensor is called the *strain-rate tensor* (denoted by symbol $\underline{\mathrm{S}}$):

$$\underline{S} = \frac{(\nabla V) + (\nabla V)^T}{2}. \tag{4.41}$$

In a Cartesian coordinate system, the strain-rate tensor is expressed as:

$$\underline{S} = S_{11}\hat{e}_1\hat{e}_1 + S_{12}\hat{e}_1\hat{e}_2 + S_{13}\hat{e}_1\hat{e}_3 + S_{21}\hat{e}_2\hat{e}_1 + S_{22}\hat{e}_2\hat{e}_2 + S_{23}\hat{e}_2\hat{e}_3 + S_{31}\hat{e}_3\hat{e}_1 + S_{32}\hat{e}_3\hat{e}_2 + S_{33}\hat{e}_3\hat{e}_3, \tag{4.42}$$

where it can be verified that the various scalar components of the tensor are:

$$
\begin{aligned}
S_{11} &= \frac{\partial V_1}{\partial x_1}, & S_{12} &= \frac{1}{2}\left(\frac{\partial V_2}{\partial x_1} + \frac{\partial V_1}{\partial x_2}\right), & S_{13} &= \frac{1}{2}\left(\frac{\partial V_3}{\partial x_1} + \frac{\partial V_1}{\partial x_3}\right), \\
S_{21} &= \frac{1}{2}\left(\frac{\partial V_1}{\partial x_2} + \frac{\partial V_2}{\partial x_1}\right), & S_{22} &= \frac{\partial V_2}{\partial x_2}, & S_{23} &= \frac{1}{2}\left(\frac{\partial V_3}{\partial x_2} + \frac{\partial V_2}{\partial x_3}\right), \\
S_{31} &= \frac{1}{2}\left(\frac{\partial V_1}{\partial x_3} + \frac{\partial V_3}{\partial x_1}\right), & S_{32} &= \frac{1}{2}\left(\frac{\partial V_2}{\partial x_3} + \frac{\partial V_3}{\partial x_2}\right), & S_{33} &= \frac{\partial V_3}{\partial x_3}.
\end{aligned}
\tag{4.43}
$$

The components S_{11}, S_{22}, and S_{33} are collectively referred to as the *normal* strain-rate components, whereas the other six scalar components are called the *shear* strain-rate components. Clearly, the normal strain-rate components are identical to the expressions obtained in (4.28)–(4.30). Thus, the normal components S_{11}, S_{22}, and S_{33} individually represent the rates of fractional change of the edges AB, AD, and AE, respectively. Further, we observe that the sum of the normal strain-rate components ($S_{ii} = S_{11} + S_{22} + S_{33}$) equals the dilatation-rate of the fluid element ABCDEFGH (4.31). A comparison of the rhs of (4.32) with the expression of S_{12} in (4.43) shows that the shear strain-rate component S_{12} equals *half* the rate at which the projections of AB' and AD' tend to align with each other (Fig. 4.9). Similarly, the shear strain-rate component S_{23} equals *half* the rate at which the projections of AD'' and AE'' tend to align with each other. The shear strain-rate component S_{31} equals *half* the rate at which the projections of AE''' and AB''' tend to align with each other. This completes our discussion on the kinematic aspects of a deformable fluid element. Further physical significance of these strain-rate components will be discussed in Chap. 5 in the context of the internal forces that exist in a fluid continuum.

Chapter 5
Governing Equations of Fluid Motion

In Chap. 3 we identified velocity $\underline{V}(\underline{X}, t)$, density $\rho(\underline{X}, t)$, temperature $T(\underline{X}, t)$ and pressure $p(\underline{X}, t)$ as the dependent variables of the Eulerian description of a flow field. To have the ability to compute and predict the values of these dependent variables, we must know their governing equations. These governing equations must emerge from the fundamental laws and axioms of classical mechanics, namely the *law of conservation of mass*, the *Euler's axioms*, *Newton's third law of motion*, and the *laws of thermodynamics*. Furthermore, we may also need an appropriate thermodynamic state equation for the medium relating to pressure, temperature, and density of the fluid. When the particle itself is an independent variable (the Lagrangian description), the application of these laws and axioms is straightforward because all the fundamental laws and axioms in their original forms are indeed applicable to independently identified particles/systems of particles. However, in the Eulerian description, wherein the particle is not an independent variable and instead the spatial location vector ($\underline{X}$) is an independent variable, we need some more advanced concepts to implement these fundamental laws/axioms in order to extract the desired governing equations of the dependent variables of our interest.

5.1 Flux Through a Surface

The meaning of the word *flux* is *flowing out*. Mathematically, *flux* is defined as the amount of entity, like mass/momentum/energy being transported across an identified surface in a flow field per unit time. Consider the surface S shown in Fig. 5.1. We identify the two sides of this surface as Side A and Side B, as shown in the figure. Our interest is to find the flux of an entity of interest across this surface from side A to side B caused by the transporting action of the velocity field existing in the domain with respect to the reference frame in context. Such a flux is also called the *advective flux*. In Fig. 5.1, the symbol dA denotes an infinitesimal part of S with $\hat{e}_n$

© The Author(s), under exclusive license to Springer Nature Switzerland AG 2026

S. S. Sinha, *Fundamentals of Fluid Mechanics*,
https://doi.org/10.1007/978-3-032-04210-1_5

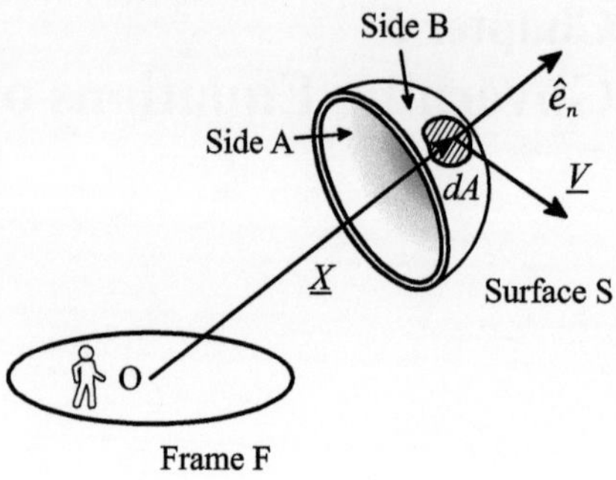

Fig. 5.1 An imaginary surface S and the flux of an entity through it

being the local unit normal vector pointing in the direction from side A to side B. The center of this infinitesimal area has its position vector as $\underline{X}$. The volume of fluid ($d\forall$) getting transported across dA, from side A to side B, per unit time, must be

$$d\forall = \left(\underline{V}(\underline{X}, t) \cdot \hat{e}_n\right) dA, \tag{5.1}$$

where $\underline{V}$ is the local instantaneous velocity vector at $\underline{X}$. Now let $\phi(\underline{X}, t)$ represent the per unit volume concentration of that entity (mass/ momentum/ energy) of interest at $\underline{X}$. Thus, the amount of the entity being transported across dA per unit of time is

$$d\mathcal{F} = \underline{\phi}(\underline{X}, t) \left(\underline{V}(\underline{X}, t) \cdot \hat{e}_n\right) dA. \tag{5.2}$$

Integrating the last equation over the entire surface results in the desired expression for the advective flux of an entity of interest across S from side A to side B.

$$\mathcal{F} = \iint_S d\mathcal{F} = \underline{\phi}(\underline{X}, t) \left(\underline{V}(\underline{X}, t) \cdot \hat{e}_n\right) dA. \tag{5.3}$$

The double-integration symbol implies integration over the entire surface S. Using this general definition of advective flux (5.3), the expression for mass flux, momentum flux, and total energy flux can be readily derived by choosing ϕ appropriately.

Setting $\phi = \rho$, which is mass per unit volume of fluid, leads to the expression of mass flux (kgs^{-1}) through S (in the direction from side A to side B):

$$\mathcal{F}_{mass} = \iint_S \rho(\underline{V} \cdot \hat{e}_n) dA. \tag{5.4}$$

For algebraic brevity, we have suppressed the repeated inclusions of the Eulerian arguments (($\underline{X}, t$)), but their presence with the relevant dependent variables is still implied. Setting $\underline{\phi} = \rho\underline{V}$, which is the momentum vector per unit volume of fluid, leads to the expression of momentum flux vector ($kgms^{-2}$) through S (in the direction from side A to side B):

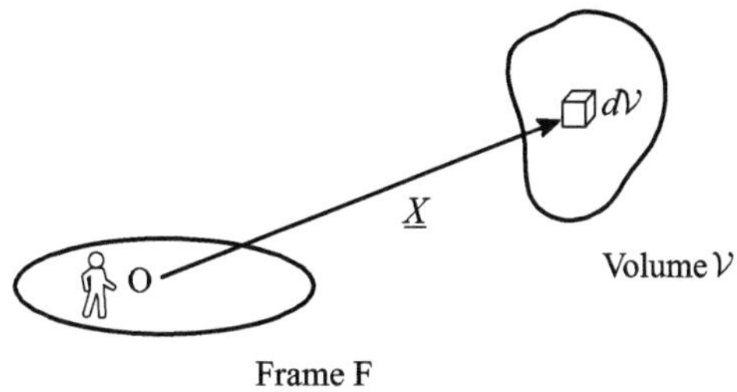

Fig. 5.2 A control volume $\mathcal{V}$. An infinitesimal volume $d\mathcal{V}$, located at $\underline{X}$, is also shown in the figure

$$\underline{\mathcal{F}}_{momentum} = \iint_S \rho \underline{V}(\underline{V} \cdot \hat{e}_n) dA. \tag{5.5}$$

Setting $\phi = \rho(\underline{V} \cdot \underline{V}/2 + e)$, which is the sum of kinetic energy and internal energy per unit volume or *total energy* per unit volume, leads to the expression of total energy flux (Js^{-1}) through S (in the direction from side A to side B):

$$\mathcal{F}_{total\, energy} = \iint_S \rho \left[\frac{\underline{V} \cdot \underline{V}}{2} + e \right] (\underline{V} \cdot \hat{e}_n) dA, \tag{5.6}$$

where *e* represents internal energy per unit mass of the fluid.

5.2 A Control Volume and the Reynolds Transport Theorem

Consider an imaginary volume enclosed by a surface $\mathcal{S}$ in a flow field (Fig. 5.2). With respect to the reference frame in context (frame F), the surface is fixed (no change in shape, size, or location). We wish to quantify the net rate (per unit time) at which an entity of interest (mass/momentum/energy) accumulates in this volume. Such a volume is often called a *control volume* in the study of fluid mechanics.

In the Eulerian description, the net rate at which an entity (with $\phi(\underline{X}, t)$ representing its concentration per unit volume of the fluid) is accumulating inside the control volume is mathematically expressed as:

$$\frac{\partial}{\partial t} \iiint_{\mathcal{V}} \phi(\underline{X}, t) d\mathcal{V}, \tag{5.7}$$

where $d\mathcal{V}$ is an infinitesimal volume which is part of the control volume $\mathcal{V}$. The triple integral symbol implies integration over the entire volume $\mathcal{V}$. All the causes that contribute to this net accumulation, we choose to categorize them into two groups $\underline{\mathcal{R}_1}$ and $\underline{\mathcal{R}_2}$ such that

$$\frac{\partial}{\partial t} \iiint_{\mathcal{V}} \underline{\phi} d\mathcal{V} = \underline{\mathcal{R}}_1 + \underline{\mathcal{R}}_2, \tag{5.8}$$

where $\underline{\mathcal{R}_1}$ and $\underline{\mathcal{R}_2}$ must be tensors of the same order as that of $\underline{\phi}$. Again, for algebraic brevity, we have suppressed the repeated inclusions of the Eulerian arguments $(\underline{X}, t)$, but their presence with the relevant dependent variables is still implied.

The first process $(\underline{\mathcal{R}_1})$ is identified as the net advective flux of the entity of interest *entering* the control volume. On the other hand, the symbol $\underline{\mathcal{R}_2}$ serves as the mathematical placeholder for all other possible causes that can contribute to the net accumulation rate of the entity of interest inside the control volume.

The advective flux represented by $\underline{\mathcal{R}_1}$ can be expressed in a general form for any arbitrary $\underline{\phi}$. Using (5.3) we express $\underline{\mathcal{R}_1}$ as:

$$\underline{\mathcal{R}_1} = - \iint_S \underline{\phi}(\underline{V} \cdot \hat{e}_n) dA. \tag{5.9}$$

This surface integral is over the entire surface, enclosing the control volume. Here, $\hat{e}_n$ represents the local outward unit normal on the surface enclosing the control volume. At every location, it points away from the control volume. The negative sign in (5.9) signifies the fact that $\underline{\mathcal{R}_1}$ denotes the net rate at which the entity is being *brought into* the control volume. It is quite possible that at certain locations, the relative orientation of the local velocity vector with the local outward normal vector is such that the relevant entity is being transported out of the control volume. The algebraic expression in (5.9) does take such variations into account.

To identify the physical causes represented by $\mathcal{R}_2$, it is often helpful to focus on the *system* of fluid mass that is currently contained inside the control volume and to imagine that suddenly *somehow the flow field is switched off*. In such an imagined situation, we ask, *what can possibly cause a change in the amount of the entity associated with this fluid mass currently residing inside the control volume?* The pertinent answer to this question is provided by the fundamental laws and axioms of classical mechanics.

In the case of the entity being mass, the relevant guidance is provided by the law of conservation of mass, which suggests that, actually, there is no means by which the fluid mass which is currently occupying the control volume can change itself. In this case, $\underline{\mathcal{R}_2}$ is a scalar and $R_2 = 0$. In the case of the entity being the momentum vector and the frame in the context being an inertial reference frame, Euler's first axiom suggests that the rate of change of momentum associated with the fluid mass currently contained inside the control volume must equal the net external force that is acting on this fluid mass. Thus,

$$\underline{R_2} = \underline{F}^{external}. \tag{5.10}$$

In case the entity is a total energy, and the frame in the context is an inertial reference frame, the first law of thermodynamics suggests:

$$R_2 = \dot{W} + \dot{Q}, \tag{5.11}$$

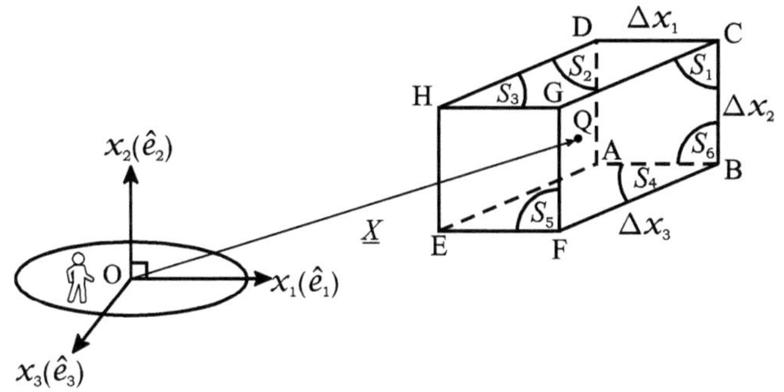

Fig. 5.3 An infinitesimal cuboidal control volume. Q is the center of the control volume located at $\underline{X}$ with coordinates (x_1,x_2,x_3)

where $\dot{W}$ represents the net rate at which external forces are doing work on the fluid mass contained inside the control volume, and $\dot{Q}$ denotes the net rate at which heat is being added to this fluid mass at this instant. In this case, $\underline{\mathcal{R}}_2$, $\dot{W}$ and $\dot{Q}$ are all scalar quantities.

Using (5.9), (5.8) is expressed as,

$$\frac{\partial}{\partial t}\iiint_{\mathcal{V}} \underline{\phi} d\mathcal{V} + \iint_{\mathcal{S}} \underline{\phi}\left(\underline{V}\cdot\hat{e}_n\right) dA = \underline{\mathcal{R}}_2. \tag{5.12}$$

This equation is called the *Reynolds transport theorem (RTT)*.

5.3 The Differential Form of RTT

Equation (5.12) is valid for a control volume of any arbitrary shape and size. In the next subsection, we plan to apply the Reynolds transport theorem to an infinitesimal-sized control volume to derive the governing equations of fluid motion. These derivations employ the Cartesian coordinate system $(Ox_1(\hat{e}_1)x_2(\hat{e}_2)x_3(\hat{e}_3))$, which is fixed to the inertial reference frame in context. Accordingly, we choose an infinitesimal control volume of cuboidal shape with the edges along $\hat{e}_1$, $\hat{e}_2$ and $\hat{e}_3$ being of small lengths Δx_1, Δx_2 and Δx_3, respectively (Fig. 5.3). The center of this control volume is located at Q, with the coordinates being (x_1,x_2,x_3).

Since the dimensions of this control volume are small, we first approximate a volume integral as the product of the integrand at the center of the CV with the volume of the CV. Thus,

$$\iiint_{\mathcal{V}} \underline{\phi} d\mathcal{V} \approx \underline{\phi}(x_1, x_2, x_3, t)\Delta x_1 \Delta x_2 \Delta x_3, \tag{5.13}$$

where $\underline{\phi}(x_1, x_2, x_3, t)$ represents the local instantaneous value of $\underline{\phi}$ at the center (Q) of the control volume.

The surface integral appearing in the advective flux term in (5.12) takes contributions from the six external surfaces of the control volume of Figure (5.3). This integral can be split into six parts corresponding to the six surfaces of the CV.

$$\begin{aligned}\iint_S \underline{\phi} \left(\underline{V} \cdot \hat{e}_n\right) dA = &\iint_{S1} \underline{\phi} \left(\underline{V} \cdot \hat{e}_n\right) dA + \iint_{S2} \underline{\phi} \left(\underline{V} \cdot \hat{e}_n\right) dA \\ &+ \iint_{S3} \underline{\phi} \left(\underline{V} \cdot \hat{e}_n\right) dA + \iint_{S4} \underline{\phi} \left(\underline{V} \cdot \hat{e}_n\right) dA \\ &+ \iint_{S5} \underline{\phi} \left(\underline{V} \cdot \hat{e}_n\right) dA + \iint_{S6} \underline{\phi} \left(\underline{V} \cdot \hat{e}_n\right) dA.\end{aligned} \tag{5.14}$$

The symbols $\mathcal{S}1$ and $\mathcal{S}2$ denote those external surfaces of the control volume which have their outward (pointing away from the control volume) unit normal vectors along the direction of $\hat{e}_1$ and $-\hat{e}_1$, respectively (see Fig. 5.3). $\mathcal{S}3$ and $\mathcal{S}4$ denote those external surfaces of the control volume which have their outward unit normal vectors along the direction of $\hat{e}_2$ and $-\hat{e}_2$, respectively. Finally, the symbols $\mathcal{S}5$ and $\mathcal{S}6$ denote those external surfaces of the control volume which have their outward unit normal vectors along the direction of $\hat{e}_3$ and $-\hat{e}_3$, respectively. Accordingly, (5.14) is expressed as:

$$\begin{aligned}\iint_S \underline{\phi} \left(\underline{V} \cdot \hat{e}_n\right) dA = &\iint_{S1} \underline{\phi} \left(\underline{V} \cdot \hat{e}_1\right) dA - \iint_{S2} \underline{\phi} \left(\underline{V} \cdot \hat{e}_1\right) dA \\ &+ \iint_{S3} \underline{\phi} \left(\underline{V} \cdot \hat{e}_2\right) dA - \iint_{S4} \underline{\phi} \left(\underline{V} \cdot \hat{e}_2\right) dA \\ &+ \iint_{S5} \underline{\phi} \left(\underline{V} \cdot \hat{e}_3\right) dA - \iint_{S6} \underline{\phi} \left(\underline{V} \cdot \hat{e}_3\right) dA.\end{aligned} \tag{5.15}$$

Further, substituting $\underline{V} \cdot \hat{e}_1 = V_1$, $\underline{V} \cdot \hat{e}_2 = V_2$, and $\underline{V} \cdot \hat{e}_3 = V_3$, we arrive at

$$\begin{aligned}\iint_S \underline{\phi} \left(\underline{V} \cdot \hat{e}_n\right) dA = &\iint_{S1} \underline{\phi} V_1 dA - \iint_{S2} \underline{\phi} V_1 dA \\ &+ \iint_{S3} \underline{\phi} V_2 dA - \iint_{S4} \underline{\phi} V_2 dA \\ &+ \iint_{S5} \underline{\phi} V_3 dA - \iint_{S6} \underline{\phi} V_3 dA.\end{aligned} \tag{5.16}$$

Since the dimensions (Δx_1, Δx_2, and Δx_3) of our cuboidal control volume are small, the areas of the surfaces $\mathcal{S}1$ to $\mathcal{S}6$, too, are small. Thus, the surface integrals appearing on the rhs of (5.16) can be approximated as the product of the value of the function to be integrated at the centroid of the respective surface and the area of that surface. This leads to the individual terms of (5.16) being expressed as

$$
\begin{aligned}
\iint_{S1} \underline{\phi} V_1 dA &\approx \underline{\phi}\left(x_1 + \frac{\Delta x_1}{2}, x_2, x_3\right) V_1\left(x_1 + \frac{\Delta x_1}{2}, x_2, x_3\right) \Delta x_2 \Delta x_3, \\
\iint_{S2} \underline{\phi} V_1 dA &\approx \underline{\phi}\left(x_1 - \frac{\Delta x_1}{2}, x_2, x_3\right) V_1\left(x_1 - \frac{\Delta x_1}{2}, x_2, x_3\right) \Delta x_2 \Delta x_3, \\
\iint_{S3} \underline{\phi} V_2 dA &\approx \underline{\phi}\left(x_1, x_2 + \frac{\Delta x_2}{2}, x_3\right) V_2\left(x_1, x_2 + \frac{\Delta x_2}{2}, x_3\right) \Delta x_3 \Delta x_1, \\
\iint_{S4} \underline{\phi} V_2 dA &\approx \underline{\phi}\left(x_1, x_2 - \frac{\Delta x_2}{2}, x_3\right) V_2\left(x_1, x_2 - \frac{\Delta x_2}{2}, x_3\right) \Delta x_3 \Delta x_1, \\
\iint_{S5} \underline{\phi} V_3 dA &\approx \underline{\phi}\left(x_1, x_2, x_3 + \frac{\Delta x_3}{2}\right) V_3\left(x_1, x_2, x_3 + \frac{\Delta x_3}{2}\right) \Delta x_1 \Delta x_2, \\
\iint_{S6} \underline{\phi} V_3 dA &\approx \underline{\phi}\left(x_1, x_2, x_3 - \frac{\Delta x_3}{2}\right) V_3\left(x_1, x_2, x_3 - \frac{\Delta x_3}{2}\right) \Delta x_1 \Delta x_2. \qquad (5.17)
\end{aligned}
$$

In (5.17) $\underline{\phi}$ and $\underline{V}$ are functions of time (t), as well. However, t has not been included in the arguments of these variables merely for the sake of algebraic brevity. To achieve further simplification of (5.15), we express the product $\left(\underline{\phi} V_1\right)$ at the face centers using Taylor series approximations, with the center of the control volume being the base location. For example,

$$
\begin{aligned}
\left(\underline{\phi} V_1\right)\Big|_{\left(x_1 + \frac{\Delta x_1}{2}, x_2, x_3\right)} &= \left(\underline{\phi} V_1\right)\Big|_{(x_1, x_2, x_3)} + \left(\frac{\Delta x_1}{2}\right) \frac{\partial \left(\underline{\phi} V_1\right)}{\partial x_1}\Bigg|_{(x_1, x_2, x_3)} \\
&+ \frac{1}{2!}\left(\frac{\Delta x_1}{2}\right)^2 \frac{\partial^2 \left(\underline{\phi} V_1\right)}{\partial x_1^2}\Bigg|_{(x_1, x_2, x_3)} + H.O.T., \qquad (5.18)
\end{aligned}
$$

where H.O.T. means the higher order terms, and (a, b, c) in $\Big|_{(a,b,c)}$ are the coordinates of the location where the relevant variable or its derivative is being computed. Using such expansions for the variables involved in $\iint_{S1} \underline{\phi} V_1 dA - \iint_{S2} \underline{\phi} V_1 dA$, leads to:

$$
\begin{aligned}
\iint_{S1} \underline{\phi} V_1 dA - \iint_{S2} \underline{\phi} V_1 dA = [2 &\left(\frac{\Delta x_1}{2}\right) \frac{\partial \left(\underline{\phi} V_1\right)}{\partial x_1} \\
&+ \frac{2}{3!}\left(\frac{\Delta x_1}{2}\right)^3 \frac{\partial^3 \left(\underline{\phi} V_1\right)}{\partial x_1^3} + H.O.T.] \Delta x_2 \Delta x_3,
\end{aligned} \qquad (5.19)
$$

where all derivatives are at the center (x_1,x_2,x_3) of the control volume ABCDEFGH in Fig. 5.3. Since Δx_1 is small, we truncate this Taylor series up to first order. This results into the following approximation:

$$\iint_{S1} \underline{\phi} V_1 dA - \iint_{S2} \underline{\phi} V_1 dA \approx \Delta x_1 \frac{\partial \left(\underline{\phi} V_1\right)}{\partial x_1} \Delta x_2 \Delta x_3. \tag{5.20}$$

Following a similar set of steps, we arrive at the approximation:

$$\iint_{S3} \underline{\phi} V_2 dA - \iint_{S4} \underline{\phi} V_2 dA \approx \Delta x_2 \frac{\partial \left(\underline{\phi} V_2\right)}{\partial x_2} \Delta x_3 \Delta x_1,$$
$$\iint_{S5} \underline{\phi} V_3 dA - \iint_{S6} \underline{\phi} V_3 dA \approx \Delta x_3 \frac{\partial \left(\underline{\phi} V_3\right)}{\partial x_3} \Delta x_1 \Delta x_2. \tag{5.21}$$

Substituting (5.20) and (5.21) successively in (5.17) and then in (5.16) leads to the following simplified expression for the net advective flux entering the control volume, which is shown in Fig. 5.3:

$$\iint_{S} \underline{\phi} \left(\underline{V} \cdot \hat{e}_n\right) dA \approx \left[\frac{\partial \left(\underline{\phi} V_1\right)}{\partial x_1} + \frac{\partial \left(\underline{\phi} V_2\right)}{\partial x_2} + \frac{\partial \left(\underline{\phi} V_3\right)}{\partial x_3}\right] \Delta x_1 \Delta x_2 \Delta x_3. \tag{5.22}$$

Using (5.22) along with (5.13), (5.12) is expressed as:

$$\frac{\partial \left(\underline{\phi} \Delta x_1 \Delta x_2 \Delta x_3\right)}{\partial t} + \left[\frac{\partial \left(\underline{\phi} V_1\right)}{\partial x_1} + \frac{\partial \left(\underline{\phi} V_2\right)}{\partial x_2} + \frac{\partial \left(\underline{\phi} V_3\right)}{\partial x_3}\right] \Delta x_1 \Delta x_2 \Delta x_3 \approx \underline{\mathcal{R}}_2. \tag{5.23}$$

Since the dimensions of the control volume in Fig. 5.3 do not change with time, $\Delta x_1 \Delta x_2 \Delta x_3$ does not vary with time. Thus, (5.23) is cast as:

$$\Delta x_1 \Delta x_2 \Delta x_3 \frac{\partial \underline{\phi}}{\partial t} + \left[\frac{\partial \left(\underline{\phi} V_1\right)}{\partial x_1} + \frac{\partial \left(\underline{\phi} V_2\right)}{\partial x_2} + \frac{\partial \left(\underline{\phi} V_3\right)}{\partial x_3}\right] \Delta x_1 \Delta x_2 \Delta x_3 \approx \underline{\mathcal{R}}_2, \tag{5.24}$$

which is then re-arranged as:

$$\frac{\partial \underline{\phi}}{\partial t} + \left[\frac{\partial \left(\underline{\phi} V_1\right)}{\partial x_1} + \frac{\partial \left(\underline{\phi} V_2\right)}{\partial x_2} + \frac{\partial \left(\underline{\phi} V_3\right)}{\partial x_3}\right] \approx \frac{\underline{\mathcal{R}}_2}{\Delta x_1 \Delta x_2 \Delta x_3}. \tag{5.25}$$

This approximate equation becomes an exact equality in the limit of the edge lengths of the control volume approaching zero:

$$\frac{\partial \underline{\phi}}{\partial t} + \left[\frac{\partial \left(\underline{\phi} V_1 \right)}{\partial x_1} + \frac{\partial \left(\underline{\phi} V_2 \right)}{\partial x_2} + \frac{\partial \left(\underline{\phi} V_3 \right)}{\partial x_3} \right] = \frac{\mathcal{R}_2}{\Delta x_1 \Delta x_2 \Delta x_3}. \tag{5.26}$$

Equation (5.26) is *the differential form of the RTT* expressed in a frame-fixed Cartesian coordinate system.

5.4 The Governing Equation of ρ

Now, let us choose the entity of interest to be mass. Accordingly, $\underline{\phi}$ in (5.26) is interpreted as mass per unit volume, which is the local density (ρ) of the fluid. Further, the law of conservation of mass implies $\mathcal{R}_2 = 0$. Thus, (5.26) simplifies to

$$\frac{\partial \rho}{\partial t} + \left[\frac{\partial (\rho V_1)}{\partial x_1} + \frac{\partial (\rho V_2)}{\partial x_2} + \frac{\partial (\rho V_3)}{\partial x_3} \right] = 0, \tag{5.27}$$

which can alternatively be presented as:

$$\frac{\partial \rho}{\partial t} + V_1 \frac{\partial \rho}{\partial x_1} + V_2 \frac{\partial \rho}{\partial x_3} + V_3 \frac{\partial \rho}{\partial x_3} = -\rho \left[\frac{\partial V_1}{\partial x_1} + \frac{\partial V_2}{\partial x_2} + \frac{\partial V_3}{\partial x_3} \right]. \tag{5.28}$$

Further, using Einstein's summation rule introduced in Chap. 1, (5.28) is expressed in a compact manner as:

$$\frac{\partial \rho}{\partial t} + V_k \frac{\partial \rho}{\partial x_k} = -\rho \frac{\partial V_k}{\partial x_k}. \tag{5.29}$$

Equation (5.27) (or its alternate form such as (5.28) or (5.29)) is called the *continuity equation*. Clearly, the continuity equation is a partial differential equation involving partial derivatives of density, as well as that of velocity components with respect to time (t) and the spatial coordinates x_1, x_2 and x_3. Using the material derivative operator (3.12), (5.29) is expressed as:

$$\frac{D\rho}{Dt} = -\rho \frac{\partial V_k}{\partial x_k}. \tag{5.30}$$

The specific form of the continuity equation as presented in (5.29) with the material derivative of ρ appearing on the left-hand side is commonly referred to as the *transport equation of* ρ. It provides information about the rate of change in density following the motion of the local fluid particle.

5.5 The Governing Equation of $\underline{V}$

To derive the governing equation of the velocity vector $\underline{V}$, we choose the entity of interest to be the momentum vector, and accordingly $\underline{\phi}$ must become the momentum vector per unit volume: $\underline{\phi} = \rho \underline{V}$. Further, following Euler's first axiom, $\underline{\mathcal{R}_2}$ must equal the net external force vector acting on the mass which is currently occupying the control volume in Fig. 5.3. This fluid mass is a fluid element having the same shape as the control volume that it is currently occupying. We denote the vertices of this fluid element as A', B', C', D', E', F', G', and H'. These vertices at the current instant coincide with the vertices A, B, C, D, E, F, G, and H of the control volume. We denote this net external force on this fluid element by $\underline{F}^{external}$. Accordingly, the differential form of RTT (5.26) takes the following form:

$$\frac{\partial (\rho \underline{V})}{\partial t} + \left[\frac{\partial (\rho \underline{V} V_1)}{\partial x_1} + \frac{\partial (\rho \underline{V} V_2)}{\partial x_2} + \frac{\partial (\rho \underline{V} V_3)}{\partial x_3} \right] = \lim_{\Delta x_1,\, \Delta x_2,\, \Delta x_3 \to 0} \frac{\underline{\mathcal{F}}^{external}}{\Delta x_1 \Delta x_2 \Delta x_3}. \tag{5.31}$$

Expressing the vector $\underline{V}$ as $V_1 \hat{e}_1 + V_2 \hat{e}_2 + V_3 \hat{e}_3$ and using the fact that the unit vectors $\hat{e}_1, \hat{e}_2, \hat{e}_3$ are frame-fixed (thus their spatial derivatives vanish), it can be verified that the vector equation (5.31) can alternatively be expressed as a set of three scalar equations:

$$\frac{\partial (\rho V_1)}{\partial t} + \left[\frac{\partial (\rho V_1 V_1)}{\partial x_1} + \frac{\partial (\rho V_1 V_2)}{\partial x_2} + \frac{\partial (\rho V_1 V_3)}{\partial x_3} \right] = \frac{F_1^{external}}{\Delta x_1 \Delta x_2 \Delta x_3}, \tag{5.32}$$

$$\frac{\partial (\rho V_2)}{\partial t} + \left[\frac{\partial (\rho V_2 V_1)}{\partial x_1} + \frac{\partial (\rho V_2 V_2)}{\partial x_2} + \frac{\partial (\rho V_2 V_3)}{\partial x_3} \right] = \frac{F_2^{external}}{\Delta x_1 \Delta x_2 \Delta x_3}, \tag{5.33}$$

$$\frac{\partial (\rho V_3)}{\partial t} + \left[\frac{\partial (\rho V_3 V_1)}{\partial x_1} + \frac{\partial (\rho V_3 V_2)}{\partial x_2} + \frac{\partial (\rho V_3 V_3)}{\partial x_3} \right] = \frac{F_3^{external}}{\Delta x_1 \Delta x_2 \Delta x_3}, \tag{5.34}$$

where $F_1^{external}$, $F_2^{external}$, and $F_3^{external}$ are the three scalar components of the net external force vector in our working coordinate system.

The single net external force vector $\underline{F}^{external}$ is an equivalent representation of the algebraic sum of all the external forces acting on the fluid element $A'B'C'D'E'F'G'H'$.

These forces that act on this fluid mass can be categorized into two groups: (i) body forces and (ii) surface forces. Body forces are those external forces acting on every fluid particle that is part of the fluid element $A'B'C'D'E'F'G'H'$. On the other hand, surface forces are those external forces that act only on those fluid particles which are currently lying on the external surface of the fluid element $A'B'C'D'E'F'G'H'$. These surface forces are exerted by the fluid particles of the neighboring fluid mass or by the particles of a solid surface with which the considered control volume is in contact. In general, both the body forces and the surface forces are distributed force systems. While the body forces are distributed over the entire volume of the fluid element $A'B'C'D'E'F'G'H'$, the surface forces are distributed over the enclosing external surface of the fluid element. Let $\underline{\mathcal{F}}^{externalbody}$ and $\underline{\mathcal{F}}^{externalsurface}$ represent the equivalent force vectors of the respective distributed force systems, then

$$\underline{\mathcal{F}}^{external} = \underline{\mathcal{F}}^{externalbody} + \underline{\mathcal{F}}^{externalsurface}. \tag{5.35}$$

A typical example of a body force system is the gravitational force system. In applications where the gravitational acceleration vector $\underline{g}$ acting on all the particles of the system is identical both in terms of magnitude and direction, we have

$$\underline{\mathcal{F}}^{externalbody} = \Delta m \underline{g}, \tag{5.36}$$

where Δm represents the mass of the fluid element $A'B'C'D'E'F'G'H'$. This mass can be expressed as:

$$\Delta m \approx \rho \Delta x_1 \Delta x_2 \Delta x_3. \tag{5.37}$$

where ρ is the instantaneous density at the center of the fluid element $A'B'C'D'E'F'G'H'$ (x_1, x_2, x_3). In turn, this implies

$$\underline{\mathcal{F}}^{externalbody} \approx \rho \Delta x_1 \Delta x_2 \Delta x_3 \underline{g}. \tag{5.38}$$

Thus, in our Cartesian coordinate system,

$$\underline{\mathcal{F}}^{externalbody} \approx \rho \Delta x_1 \Delta x_2 \Delta x_3 \left[g_1 \hat{e}_1 + g_2 \hat{e}_2 + g_3 \hat{e}_3\right], \tag{5.39}$$

where g_1, g_2 and g_3 are the scalar components of the vector $\underline{g}$.

5.5.1 The Stress Tensor

We define a new second-order tensor called the *stress* tensor to represent the surface forces mathematically. We denote this quantity by the symbol $\underline{\sigma}$. In the Eulerian description of fluid motion, the stress tensor, too, is cast as an Eulerian variable. It

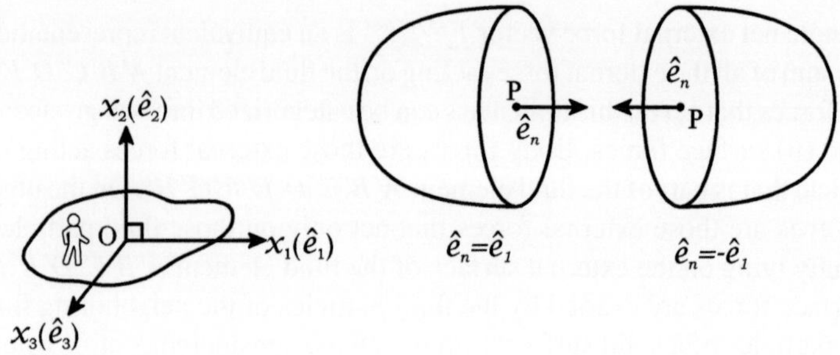

Fig. 5.4 Two parts of the fluid bulk created by an imaginary plane which is perpendicular to $\hat{e}_1$. The plane passed through the location P. The two parts appear separated and displaced only for the purpose of illustration

depends on spatial location and time, $\underline{\sigma}(\underline{X}, t)$. In our working Cartesian coordinate system, $\underline{\sigma}$ is expressed as:

$$\underline{\sigma} = \sigma_{11}\hat{e}_1\hat{e}_1 + \sigma_{12}\hat{e}_1\hat{e}_2 + \sigma_{13}\hat{e}_1\hat{e}_3 + \sigma_{21}\hat{e}_2\hat{e}_1 + \sigma_{22}\hat{e}_2\hat{e}_2 + \sigma_{23}\hat{e}_2\hat{e}_3 + \sigma_{31}\hat{e}_3\hat{e}_1 + \sigma_{32}\hat{e}_3\hat{e}_2 + \sigma_{33}\hat{e}_3\hat{e}_3 = \sigma_{ij}\hat{e}_i\hat{e}_j. \quad (5.40)$$

where $\sigma_{ij}(x_1, x_2, x_3, t)$ represents the (ij)th components of the instantaneous local stress tensor.

To understand the relationship between the surface forces and the various stress components, let us consider a bulk of fluid, as shown in Fig. 5.4. Further, we consider an imaginary plane perpendicular to the direction of the unit vector $\hat{e}_1$ which passes through location P which has its coordinates as (x_1, x_2, x_3) (Fig. 5.4). This plane divides the fluid bulk into two parts. We focus on the left part and identify this mass of the fluid as our system. The outward normal unit vector for this exposed surface of the left bulk of the fluid is $\hat{e}_1$. We now wish to identify the surface forces exerted by the right part of the fluid bulk on the left part. While these forces would have been the internal forces for the entire fluid bulk, these must be accounted as *external surface forces* when the left bulk of the fluid mass is being considered as a free body. In general, these surface forces make a distributed force system. However, on a small area ΔA with its centroid at location P (x_1, x_2, x_3), the small contribution of this force system can be represented by a single equivalent force vector, $\Delta\underline{F}$ (see Fig. 5.5). This force vector is expressed in the Cartesian coordinate system as:

$$\Delta\underline{F} = \Delta F_1\hat{e}_1 + \Delta F_2\hat{e}_2 + \Delta F_3\hat{e}_3. \quad (5.41)$$

The three scalar Cartesian components σ_{11}, σ_{12} and σ_{13} of the stress tensor at location P with coordinates x_1, x_2, x_3 are related to ΔF_1, ΔF_2 and ΔF_3 as:

$$\sigma_{11} = \lim_{\Delta A \to 0} \frac{\Delta F_1}{\Delta A}, \; \sigma_{12} = \lim_{\Delta A \to 0} \frac{\Delta F_2}{\Delta A}, \; \sigma_{13} = \lim_{\Delta A \to 0} \frac{\Delta F_3}{\Delta A}. \quad (5.42)$$

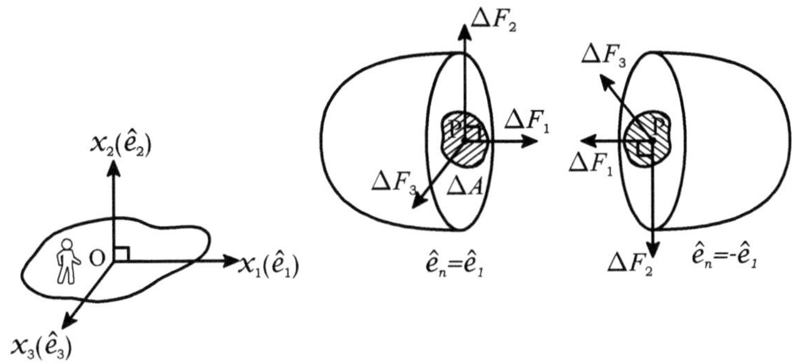

Fig. 5.5 Surface forces on small shaded areas with their outward unit normal vectors along the $\hat{e}_1$ and $-\hat{e}_1$ directions. The location P with its coordinates (x_1, x_2, x_3) is the centroid of ΔA

Clearly, the first index of a stress component is the same as the index of the unit vector normal to ΔA. On the other hand, the second index of that stress component is the same as the index of the corresponding force component.

The sign ($\pm$) of a scalar component of the stress tensor is determined based on both (i) the direction of the force components and (ii) the direction of the outward normal vector of ΔA:

1. The sign of a scalar component σ_{ij} of the stress tensor is positive if the outward normal is along the $+\hat{e}_i$ direction and the scalar force component ΔF_j is positive.
2. The sign of a scalar component σ_{ij} is positive if the outward normal is along the $-\hat{e}_i$ direction and the scalar force component ΔF_j is negative.
3. The sign of a scalar component σ_{ij} of the stress tensor is negative if the outward normal is along $+\hat{e}_i$ direction and the scalar force component ΔF_j is negative.
4. The sign of a scalar component σ_{ij} of the stress tensor is negative if the outward normal is along the $-\hat{e}_i$ direction and the scalar force component ΔF_j is positive.

Following this set of rules, it can be verified that the magnitudes and the signs of the three components of the stress tensor explained in (5.42) would turn out to be identical even if we choose to work with the right part of the fluid bulk shown in Fig. 5.5 as the free body of our analysis. Following Newton's third law of motion, the external surface force vector acting on the right bulk of the fluid must be equal, and it must be oriented in the opposite direction as compared to the corresponding external force on the left part. However, the small area ΔA through which this force acts on the system on the right-hand side has its outward normal in $-\hat{e}_1$ direction (Fig. 5.5). Thus, the signs of the force components along with the sign of the outward unit normal vector will consistently adjust with each other, resulting in the same signs and the magnitudes of all the components of the stress tensor, as we had when we used the left part of the bulk for our analysis in (5.41) and (5.42). Thus, we say that the stress tensor itself (and consequently all its scalar components with reference to the working coordinate system) is an Eulerian variable that depends only on time and location. There is no reference to any surface required for the description of stress

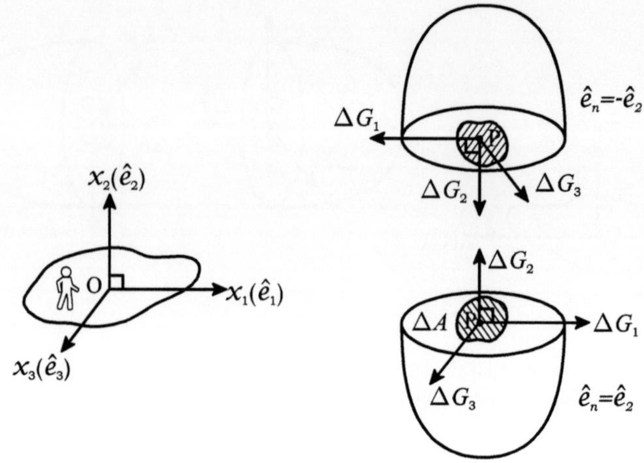

Fig. 5.6 Surface forces on small shaded areas with their outward unit normal vectors along the $\hat{e}_2$ and $-\hat{e}_2$ directions. The location P with its coordinates (x_1,x_2,x_3) is the centroid of ΔA

at a point. However, when we seek to find surface forces using the stress tensor at a point, we must first define the orientation of a surface passing through that point. The relevant components of the stress tensor can then provide information about the infinitesimal force vector acting on an infinitesimal area of that surface at that point.

The physical meanings of the other Cartesian components of the stress tensor at location P can also be explained.

$$\sigma_{21} = \lim_{\Delta A \to 0} \frac{\Delta G_1}{\Delta A}, \ \sigma_{22} = \lim_{\Delta A \to 0} \frac{\Delta G_2}{\Delta A}, \ \sigma_{23} = \lim_{\Delta A \to 0} \frac{\Delta G_3}{\Delta A}, \tag{5.43}$$

where ΔA in (5.43) is a small area with its centroid at P with its coordinates (x_1,x_2,x_3) on the exposed surface which is created by a plane which has its outward normal along $\hat{e}_2$ direction (Fig. 5.6). The symbols ΔG_1, ΔG_2, and ΔG_3 represent the Cartesian components of the surface force exerted by the surrounding fluid on the shown system through the area ΔA in the figure.

Finally, the physical meanings of σ_{31}, σ_{32}, and σ_{33} at location P can be explained as follows:

$$\sigma_{31} = \lim_{\Delta A \to 0} \frac{\Delta H_1}{\Delta A}, \ \sigma_{32} = \lim_{\Delta A \to 0} \frac{\Delta H_2}{\Delta A}, \ \sigma_{33} = \lim_{\Delta A \to 0} \frac{\Delta H_3}{\Delta A}, \tag{5.44}$$

where ΔA in (5.44) is a small area with its centroid at P with its coordinates as (x_1,x_2,x_3) on the exposed surface which is created by a plane which has its outward normal along $\hat{e}_3$ direction (Fig. 5.7). The symbols ΔH_1, ΔH_2, and ΔH_3 represent the Cartesian components of the surface force vector exerted by the surrounding fluid on the shown area ΔA in the figure.

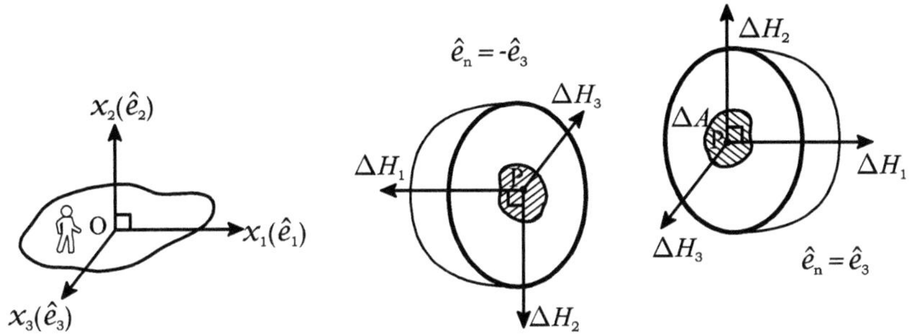

Fig. 5.7 Surface forces on small shaded areas with the outward unit normal vectors along the $\hat{e}_3$ and $-\hat{e}_3$ directions. The location P with its coordinates (x_1,x_2,x_3) is the centroid of ΔA

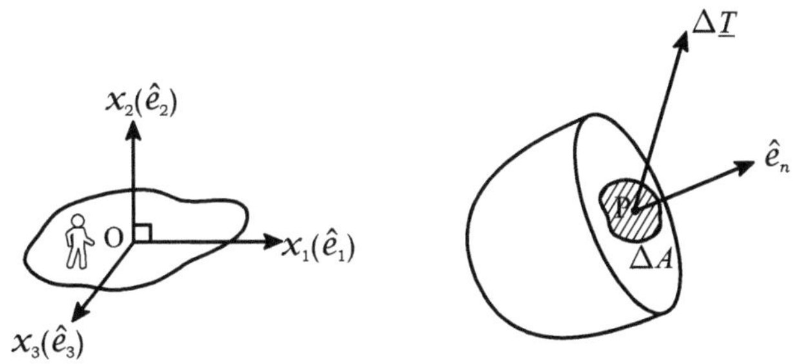

Fig. 5.8 The surface force vector displayed on a small area with its outward unit normal vector along the $\hat{e}_n$ direction. The location P is the centroid of ΔA

Those scalar components of the stress tensor which describe the force components normal to a surface are called the *normal stress* components, whereas the other components are called the *shear stress* components. Thus, in the Cartesian coordinate system, the components σ_{11}, σ_{22}, and σ_{33} are the normal stress components and σ_{12}, σ_{23}, σ_{13}, σ_{21}, σ_{32}, and σ_{31} the shear stress components.

In the foregoing discussion, our aim has been to explain the meanings of the scalar components σ_{ij} with reference to the Cartesian coordinate system $Ox_1(\hat{e}_1)x_2(\hat{e}_2)x_3(\hat{e}_3)$. Accordingly, we chose the bulk-dividing planes to be perpendicular to one of the three basis vectors of our coordinate system. However, there exists a relationship between the local surface force vector (say, $\Delta\underline{T}$), which acts on a small area ΔA that exists on an imaginary surface created using a plane having its normal oriented along an arbitrary direction $\hat{e}_n$ (Fig. 5.8).

$$\hat{e}_n \cdot \underline{\sigma} = \lim_{\Delta A \to 0} \frac{\Delta \underline{T}}{\Delta A}. \tag{5.45}$$

It can be verified that all individual relationships listed in (5.42), (5.43), and (5.44) are indeed special cases of (5.45).

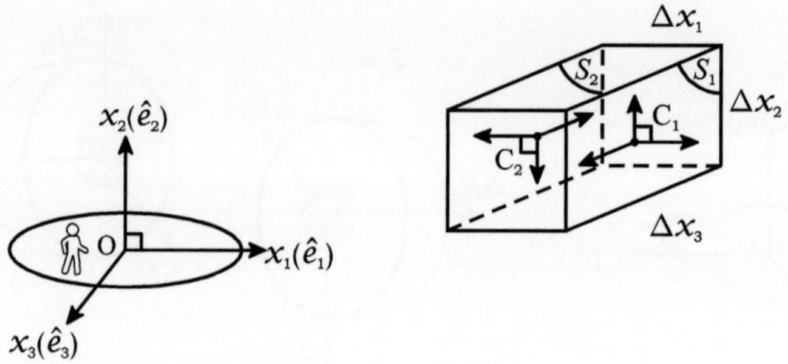

Fig. 5.9 Cartesian components of the surface force vectors acting on faces S1 and S2 on the shown fluid element. The shown forces act at the face centroids (C1 and C2)

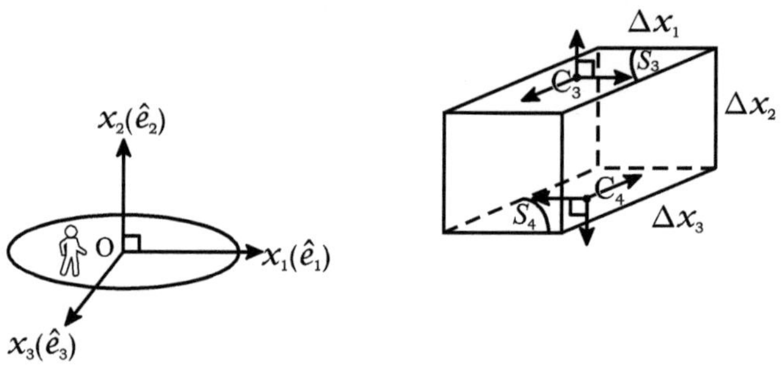

Fig. 5.10 Cartesian components of the surface force vectors acting on faces S3 and S4 on the shown fluid element. The shown forces act at the face centroids (C3 and C4)

5.5.2 *External Force on a Fluid Element*

Using the definition of the stress tensor (5.45), we now intend to find the expression for $\underline{F}^{externalsurface}$ acting on the fluid element $A'B'C'D'E'F'G'H'$ of Fig. 5.3. In Figs. 5.9, 5.10 and 5.11 we have labeled the six surfaces of the cuboidal fluid element $A'B'C'D'$–$E'F'G'H'$. We must identify the respective surface forces on each of these six surfaces. Again, the surface forces on these faces would generally be distributed. Nonetheless, these faces being small, the system of distributed forces on each face can be replaced by an equivalent force system, which has a single force vector acting at the centroid of the respective face. The centroids of the six faces S1 to S6 are labeled as C1 to C6 in Figs. 5.9, 5.10 and 5.11. The coordinates of these centroids are

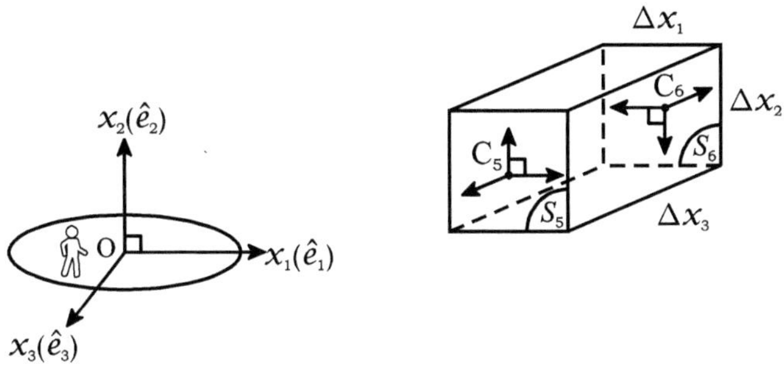

Fig. 5.11 Cartesian components of the surface force vectors acting on faces S5 and S6 on the shown fluid element. The shown forces act at the face centroids (C5 and C6)

$$
\begin{aligned}
&\text{C1} : \left(x_1 + \frac{\Delta x_1}{2}, x_2, x_3\right), \text{C2} : \left(x_1 - \frac{\Delta x_1}{2}, x_2, x_3\right), \\
&\text{C3} : \left(x_1, x_2 + \frac{\Delta x_2}{2}, x_3\right), \text{C4} : \left(x_1, x_2 - \frac{\Delta x_2}{2}, x_3\right), \\
&\text{C5} : \left(x_1, x_2, x_3 + \frac{\Delta x_3}{2}\right), \text{C6} : \left(x_1, x_2, x_3 - \frac{\Delta x_3}{2}\right).
\end{aligned} \tag{5.46}
$$

That single force vector itself is expressed as a product of the face area and the relevant component of the stress tensor at the center of the face. In Fig. 5.9 we show the components of the surface forces acting on the faces S1 and S2 of the fluid element $A'B'C'D'E'F'G'H'$. Figure 5.10 shows the components of surface forces acting on faces S3 and S4. Finally, Fig. 5.11 shows the components of surface forces acting on faces S5 and S6.

The force vectors acting on the six faces are (the expressions use index notation for algebraic brevity, and summations are implied over the index k)

$$
\begin{aligned}
&\text{Face S1:} \quad \left[\sigma_{1k}\left(x_1 + \frac{\Delta x_1}{2}, x_2, x_3\right)\right] \Delta x_2 \Delta x_3 \hat{e}_k, \\
&\text{Face S2:} \quad -\left[\sigma_{1k}\left(x_1 - \frac{\Delta x_1}{2}, x_2, x_3\right)\right] \Delta x_2 \Delta x_3 \hat{e}_k, \\
&\text{Face S3:} \quad \left[\sigma_{2k}\left(x_1, x_2 + \frac{\Delta x_2}{2}, x_3\right)\right] \Delta x_3 \Delta x_1 \hat{e}_k, \\
&\text{Face S4:} \quad -\left[\sigma_{2k}\left(x_1, x_2 - \frac{\Delta x_2}{2}, x_3\right)\right] \Delta x_3 \Delta x_1 \hat{e}_k, \\
&\text{Face S5:} \quad \left[\sigma_{3k}\left(x_1, x_2, x_3 + \frac{\Delta x_3}{2}\right)\right] \Delta x_1 \Delta x_2 \hat{e}_k, \\
&\text{Face S6:} \quad -\left[\sigma_{3k}\left(x_1, x_2, x_3 - \frac{\Delta x_3}{2}\right)\right] \Delta x_1 \Delta x_2 \hat{e}_k.
\end{aligned} \tag{5.47}
$$

Now with reference to Fig. 5.9, the net algebraic sum of the surface forces acting on S1 and S2 (denoted by symbol $\underline{F}^{external-surface1,2}$) is:

$$
\begin{aligned}
&\underline{F}^{external-surface1,2} \approx \\
&\quad \left[\sigma_{11}(x_1+\Delta x_1/2, x_2, x_3)\hat{e}_1 + \sigma_{12}(x_1+\Delta x_1/2, x_2, x_3)\hat{e}_2 \right.\\
&\quad \left. +\sigma_{13}(x_1+\Delta x_1/2, x_2, x_3)\hat{e}_3\right]\Delta x_2 \Delta x_3 \\
&\quad -\left[\sigma_{11}(x_1-\Delta x_1/2, x_2, x_3)\hat{e}_1 + \sigma_{12}(x_1-\Delta x_1/2, x_2, x_3)\hat{e}_2 \right.\\
&\quad \left. +\sigma_{13}(x_1-\Delta x_1/2, x_2, x_3)\hat{e}_3\right]\Delta x_2 \Delta x_3.
\end{aligned} \tag{5.48}
$$

With reference to Fig. 5.10, the net algebraic sum of the surface forces acting on S3 and S4 (denoted by symbol $\underline{F}^{external-surface3,4}$) is:

$$
\begin{aligned}
&\underline{F}^{external-surface3,4} \approx \\
&\quad \left[\sigma_{21}(x_1, x_2+\Delta x_2/2, x_3)\hat{e}_1 + \sigma_{22}(x_1, x_2+\Delta x_2/2, x_3)\hat{e}_2 \right.\\
&\quad \left. +\sigma_{23}(x_1, x_2+\Delta x_2/2, x_3)\hat{e}_3\right]\Delta x_3 \Delta x_1 \\
&\quad -\left[\sigma_{21}(x_1, x_2-\Delta x_2/2, x_3)\hat{e}_1 + \sigma_{22}(x_1, x_2-\Delta x_2/2, x_3)\hat{e}_2 \right.\\
&\quad \left. +\sigma_{23}(x_1, x_2-\Delta x_2/2, x_3)\hat{e}_3\right]\Delta x_3 \Delta x_1.
\end{aligned} \tag{5.49}
$$

With reference to Fig. 5.11, the net algebraic sum of the surface forces acting on S5 and S6 (denoted by symbol $\underline{F}^{external-surface5,6}$) is:

$$
\begin{aligned}
&\underline{F}^{external-surface5,6} \approx \\
&\quad \left[\sigma_{31}(x_1, x_2, x_3+\Delta x_3/2)\hat{e}_1 + \sigma_{32}(x_1, x_2, x_3+\Delta x_3/2)\hat{e}_2 \right.\\
&\quad \left. +\sigma_{33}(x_1, x_2, x_3+\Delta x_3/2)\hat{e}_3\right]\Delta x_1 \Delta x_2 \\
&\quad -\left[\sigma_{31}(x_1, x_2, x_3-\Delta x_3/2)\hat{e}_1 + \sigma_{32}(x_1, x_2, x_3-\Delta x_3/2)\hat{e}_2 \right.\\
&\quad \left. +\sigma_{33}(x_1, x_2, x_3-\Delta x_3/2)\hat{e}_3\right]\Delta x_1 \Delta x_2.
\end{aligned} \tag{5.50}
$$

To simplify the expression of (5.48)–(5.50), we express the components of the stress tensor at the face centers using appropriate Taylor series expansion up to first order. The base location for this expansion is the centroid (point Q with coordinates (x_1,x_2,x_3) of the control volume/the fluid element occupying the control volume, Fig. 5.3). Accordingly, for surfaces S1 and S2:

$$
\begin{aligned}
\sigma_{11}(x_1+\Delta x_1/2, x_2, x_3) &\approx \sigma_{11}(x_1, x_2, x_3) + \frac{\partial \sigma_{11}}{\partial x_1}\frac{\Delta x_1}{2}, \\
\sigma_{12}(x_1+\Delta x_1/2, x_2, x_3) &\approx \sigma_{12}(x_1, x_2, x_3) + \frac{\partial \sigma_{12}}{\partial x_1}\frac{\Delta x_1}{2}, \\
\sigma_{13}(x_1+\Delta x_1/2, x_2, x_3) &\approx \sigma_{13}(x_1, x_2, x_3) + \frac{\partial \sigma_{13}}{\partial x_1}\frac{\Delta x_1}{2}, \\
\sigma_{11}(x_1-\Delta x_1/2, x_2, x_3) &\approx \sigma_{11}(x_1, x_2, x_3) - \frac{\partial \sigma_{11}}{\partial x_1}\frac{\Delta x_1}{2},
\end{aligned}
$$

$$\sigma_{12}(x_1 - \Delta x_1/2, x_2, x_3) \approx \sigma_{12}(x_1, x_2, x_3) - \frac{\partial \sigma_{12}}{\partial x_1}\frac{\Delta x_1}{2},$$
$$\sigma_{13}(x_1 - \Delta x_1/2, x_2, x_3) \approx \sigma_{13}(x_1, x_2, x_3) - \frac{\partial \sigma_{13}}{\partial x_1}\frac{\Delta x_1}{2}, \tag{5.51}$$

where all the partial derivatives are at the centroid of the control volume. Using (5.51) in (5.48) leads to

$$\underline{F}^{external-surface1,2} \approx \left[\frac{\partial \sigma_{11}}{\partial x_1}\hat{e}_1 + \frac{\partial \sigma_{12}}{\partial x_1}\hat{e}_2 + \frac{\partial \sigma_{13}}{\partial x_1}\hat{e}_3\right]\Delta x_1 \Delta x_2 \Delta x_3. \tag{5.52}$$

On surfaces S3 and S4, the relevant approximations are:

$$\sigma_{21}(x_1, x_2 + \Delta x_2/2, x_3) \approx \sigma_{21}(x_1, x_2, x_3) + \frac{\partial \sigma_{21}}{\partial x_2}\frac{\Delta x_2}{2},$$
$$\sigma_{22}(x_1, x_2 + \Delta x_2/2, x_3) \approx \sigma_{22}(x_1, x_2, x_3) + \frac{\partial \sigma_{22}}{\partial x_2}\frac{\Delta x_2}{2},$$
$$\sigma_{23}(x_1, x_2 + \Delta x_2/2, x_3) \approx \sigma_{23}(x_1, x_2, x_3) + \frac{\partial \sigma_{23}}{\partial x_2}\frac{\Delta x_2}{2},$$
$$\sigma_{21}(x_1, x_2 - \Delta x_2/2, x_3) \approx \sigma_{21}(x_1, x_2, x_3) - \frac{\partial \sigma_{21}}{\partial x_2}\frac{\Delta x_2}{2},$$
$$\sigma_{22}(x_1, x_2 - \Delta x_2/2, x_3) \approx \sigma_{22}(x_1, x_2, x_3) - \frac{\partial \sigma_{22}}{\partial x_2}\frac{\Delta x_2}{2},$$
$$\sigma_{23}(x_1, x_2 - \Delta x_2/2, x_3) \approx \sigma_{23}(x_1, x_2, x_3) - \frac{\partial \sigma_{23}}{\partial x_2}\frac{\Delta x_2}{2}. \tag{5.53}$$

Using (5.53) in (5.49) leads to

$$\underline{F}^{external-surface3,4} \approx \left[\frac{\partial \sigma_{21}}{\partial x_2}\hat{e}_1 + \frac{\partial \sigma_{22}}{\partial x_2}\hat{e}_2 + \frac{\partial \sigma_{23}}{\partial x_2}\hat{e}_3\right]\Delta x_1 \Delta x_2 \Delta x_3. \tag{5.54}$$

On surfaces S5 and S6, the relevant approximations are:

$$\sigma_{31}(x_1, x_2, x_3 + \Delta x_3/2) \approx \sigma_{31}(x_1, x_2, x_3) + \frac{\partial \sigma_{31}}{\partial x_3}\frac{\Delta x_3}{2},$$
$$\sigma_{32}(x_1, x_2, x_3 + \Delta x_3/2) \approx \sigma_{32}(x_1, x_2, x_3) + \frac{\partial \sigma_{32}}{\partial x_3}\frac{\Delta x_3}{2},$$
$$\sigma_{33}(x_1, x_2, x_3 + \Delta x_3/2) \approx \sigma_{33}(x_1, x_2, x_3) + \frac{\partial \sigma_{33}}{\partial x_3}\frac{\Delta x_3}{2},$$
$$\sigma_{31}(x_1, x_2, x_3 - \Delta x_3/2) \approx \sigma_{31}(x_1, x_2, x_3) - \frac{\partial \sigma_{31}}{\partial x_3}\frac{\Delta x_3}{2},$$
$$\sigma_{32}(x_1, x_2, x_3 - \Delta x_3/2) \approx \sigma_{32}(x_1, x_2, x_3) - \frac{\partial \sigma_{32}}{\partial x_3}\frac{\Delta x_3}{2},$$

$$\sigma_{33}(x_1, x_2, x_3 - \Delta x_3/2) \approx \sigma_{33}(x_1, x_2, x_3) - \frac{\partial \sigma_{33}}{\partial x_3}\frac{\Delta x_3}{2}. \tag{5.55}$$

Using (5.55) in (5.50) leads to

$$\underline{F}^{external-surface3,4} \approx \left[\frac{\partial \sigma_{31}}{\partial x_3}\hat{e}_1 + \frac{\partial \sigma_{32}}{\partial x_3}\hat{e}_2 + \frac{\partial \sigma_{33}}{\partial x_3}\hat{e}_3\right] \Delta x_1 \Delta x_2 \Delta x_3. \tag{5.56}$$

Assembling the contributions from the six faces (5.52), (5.54), and (5.56), we arrive at the expression of $\underline{F}^{externalsurface}$ (as required in 5.35):

$$\begin{aligned}
\underline{F}^{externalsurface} &= \underline{F}^{external-surface1,2} + \underline{F}^{externalsurface3,4} \\
&+ \underline{F}^{external-surface5,6}, \\
&\approx \left[\frac{\partial \sigma_{11}}{\partial x_1} + \frac{\partial \sigma_{21}}{\partial x_2} + \frac{\partial \sigma_{31}}{\partial x_3}\right] \Delta x_1 \Delta x_2 \Delta x_3 \hat{e}_1 + \\
&\left[\frac{\partial \sigma_{12}}{\partial x_1} + \frac{\partial \sigma_{22}}{\partial x_2} + \frac{\partial \sigma_{32}}{\partial x_3}\right] \Delta x_1 \Delta x_2 \Delta x_3 \hat{e}_2 + \\
&\left[\frac{\partial \sigma_{13}}{\partial x_1} + \frac{\partial \sigma_{23}}{\partial x_2} + \frac{\partial \sigma_{33}}{\partial x_3}\right] \Delta x_1 \Delta x_2 \Delta x_3 \hat{e}_3.
\end{aligned} \tag{5.57}$$

With the expressions for the $\underline{F}^{externalsurface}$ (5.57) and $\underline{F}^{externalbody}$ (5.39) available now, the expression for $\underline{F}^{external}$ (5.35) is:

$$\begin{aligned}
\underline{\mathcal{F}}^{external} &= \underline{\mathcal{F}}^{externalbody} + \underline{\mathcal{F}}^{externalsurface} \\
&\approx \rho \left[g_1 \hat{e} + g_2 \hat{e}_2 + g_3 \hat{e}_3\right] \Delta x_1 \Delta x_2 \Delta x_3 \\
&+ \left[\frac{\partial \sigma_{11}}{\partial x_1} + \frac{\partial \sigma_{21}}{\partial x_2} + \frac{\partial \sigma_{31}}{\partial x_3}\right] \Delta x_1 \Delta x_2 \Delta x_3 \hat{e}_1 \\
&+ \left[\frac{\partial \sigma_{12}}{\partial x_1} + \frac{\partial \sigma_{22}}{\partial x_2} + \frac{\partial \sigma_{32}}{\partial x_3}\right] \Delta x_1 \Delta x_2 \Delta x_3 \hat{e}_2 \\
&+ \left[\frac{\partial \sigma_{13}}{\partial x_1} + \frac{\partial \sigma_{23}}{\partial x_2} + \frac{\partial \sigma_{33}}{\partial x_3}\right] \Delta x_1 \Delta x_2 \Delta x_3 \hat{e}_3.
\end{aligned} \tag{5.58}$$

Selecting the appropriate scalar components of the vector equation (5.58) and using those in the governing equations of V_1 (5.32), V_2 (5.33), and V_3 (5.34), these equations are now expressed as:

$$\begin{aligned}
&\frac{\partial (\rho V_1)}{\partial t} + \left[\frac{\partial (\rho V_1 V_1)}{\partial x_1} + \frac{\partial (\rho V_1 V_2)}{\partial x_2} + \frac{\partial (\rho V_1 V_3)}{\partial x_3}\right] = \\
&\qquad \rho g_1 + \frac{\partial \sigma_{11}}{\partial x_1} + \frac{\partial \sigma_{21}}{\partial x_2} + \frac{\partial \sigma_{31}}{\partial x_3},
\end{aligned} \tag{5.59}$$

$$\frac{\partial (\rho V_2)}{\partial t} + \left[\frac{\partial (\rho V_2 V_1)}{\partial x_1} + \frac{\partial (\rho V_2 V_2)}{\partial x_2} + \frac{\partial (\rho V_2 V_3)}{\partial x_3}\right] = \rho g_2 + \frac{\partial \sigma_{12}}{\partial x_1} + \frac{\partial \sigma_{22}}{\partial x_2} + \frac{\partial \sigma_{32}}{\partial x_3}, \tag{5.60}$$

$$\frac{\partial (\rho V_3)}{\partial t} + \left[\frac{\partial (\rho V_3 V_1)}{\partial x_1} + \frac{\partial (\rho V_3 V_2)}{\partial x_2} + \frac{\partial (\rho V_3 V_3)}{\partial x_3}\right] = \rho g_3 + \frac{\partial \sigma_{13}}{\partial x_1} + \frac{\partial \sigma_{23}}{\partial x_2} + \frac{\partial \sigma_{33}}{\partial x_3}. \tag{5.61}$$

Using the index notation, (5.59), (5.60), and (5.61) can be represented using one single scalar equation with i as the free index.

$$\frac{\partial (\rho V_i)}{\partial t} + \frac{\partial \left(\rho V_j V_i\right)}{\partial x_j} = \rho g_i + \frac{\partial \sigma_{ji}}{\partial x_j}. \tag{5.62}$$

In each term of (5.62), there is exactly one free index. Moreover, in each term, the free index is i. Selecting $i = 1$ or 2 or 3 at a time leads to the three independent scalar equations (5.59), (5.60), and (5.61).

Further, taking the product of V_i with the governing equation of density (5.29) and subtracting the resulting equation from (5.62) results in a form of the governing equation of V_i, which can be readily cast in terms of the material derivative operator.

$$\frac{DV_i}{Dt} = g_i + \frac{1}{\rho}\frac{\partial \sigma_{ji}}{\partial x_j}. \tag{5.63}$$

We refer to (5.63) as the transport equation of the ith component of the velocity vector. It provides information about the rate of change in V_i (or the ith component of the acceleration vector) following the motion of a local fluid particle. The right side of this equation represents the net external force per unit mass that acts on the local fluid particle. Like the transport equation of ρ (5.29), the transport equation of V_i involves several partial derivatives. However, there is a difference: unlike the transport equation of density, which depends only on the primary flow variables of the Eulerian description of motion (ρ, V_i's), the transport equation of V_i involves additional variables—several scalar components of the local stress tensor ($\underline{\sigma}+$).

5.6 The Governing Equation of T

To derive the governing equation of temperature (T) we first choose the entity of interest to be the total energy (the sum of internal energy and the kinetic energy), and accordingly $\underline{\phi}$ in the differential form of RTT (5.26) must be the sum of internal energy and the kinetic energy per unit volume of the fluid. $\underline{\phi} = \rho\left(e + \frac{1}{2}\underline{V} \cdot \underline{V}\right)$, where e represents the local instantaneous internal energy per unit mass of the fluid. Further, following the first law of thermodynamics being applied to the fluid element $A'B'C'D'E'F'G'H'$ in Fig. 5.3, $\underline{\mathcal{R}_2}$ in (5.26) must equal the net rate at which the external forces are doing work on the fluid element ($\dot{W}$) plus the net rate at which heat is being added to the fluid element ($\dot{Q}$). This results in the following scalar equation:

$$\frac{\partial\left[\rho\left(e + \frac{1}{2}V_k V_k\right)\right]}{\partial t} + \frac{\partial\left\{\left[\rho\left(e + \frac{1}{2}V_k V_k\right)\right] V_j\right\}}{\partial x_j} = \frac{\dot{W} + \dot{Q}}{\Delta x_1 \Delta x_2 \Delta x_3}, \tag{5.64}$$

where a repeated index implies summation.

To find the mathematical expressions for $\dot{W}$, we re-consider all the external surface forces and the external body forces acting on the fluid element $A'B'C'D'E'F'G'H'$ (Figs. 5.9, 5.10, and 5.11). $\dot{W}$ must be the summation of the dot product of these individual forces and the velocity vectors of the respective fluid particles, which are present at the points of application of these forces. We express the individual contributions of the body forces and the surface forces as:

$$\dot{W} = \dot{W}^{bodyforce} + \dot{W}^{surfaceforce}, \tag{5.65}$$

where

$$\begin{aligned} \dot{W}^{bodyforce} &= \left[\underline{F}^{externalbody}\right] \cdot \underline{V}, \\ &\approx \left[\rho \underline{g} \Delta x_1 \Delta x_2 \Delta x_3\right] \cdot \underline{V}, \\ &\approx \left[\rho g_k \Delta x_1 \Delta x_2 \Delta x_3\right] V_k, \end{aligned} \tag{5.66}$$

where g_k and V_k represent the kth component of the acceleration-due-to-gravity vector and the instantaneous velocity vector at the centroid of the fluid element $A'B'C'D'E'F'G'H'$ (x_1, x_2, x_3). The expression $\underline{F}^{externalbody}$ is the same as what we derived earlier in (5.39).

The expression for $\dot{W}^{surfaceforce}$ takes contributions from the three pairs of opposite faces of the cuboidal fluid element shown in Figs. 5.9, 5.10, and 5.11:

$$\dot{W}^{surfaceforce} = \dot{W}^{surface-force1,2} + \dot{W}^{surface-force3,4} + \dot{W}^{surface-force5,6}, \tag{5.67}$$

where $\dot{W}^{surface-force1,2}$ represents the net rate of work done on the fluid element due to the external surface forces acting on faces 1 and 2 of the fluid element (Fig. 5.9). Similar meanings are associated with the symbols $\dot{W}^{surface-force3,4}$ and

$\dot{W}^{surface-force5,6}$. The net rate at which respective surface forces are doing work on faces 1 and 2 is expressed as a combination.

$$\begin{aligned}\dot{W}^{surface-force1,2} &\approx \\ &\left[\sigma_{1j}(x_1+\Delta x_1/2, x_2, x_3)\right]\Delta x_2 \Delta x_3 V_j(x_1+\Delta x_1/2, x_2, x_3) \\ &-\left[\sigma_{1j}(x_1-\Delta x_1/2, x_2, x_3)\right]\Delta x_2 \Delta x_3 V_j(x_1-\Delta x_1/2, x_2, x_3).\end{aligned} \tag{5.68}$$

The negative sign in front of the second term in (5.68) exists because, on the face S_2 of the fluid element, the directions of the three force components are oriented in the opposite sense as compared to the respective components of the velocity vectors at the face centroid. Thus, the dot product between the local force vector and the local velocity vector would result in that negative sign.

Similarly, $\dot{W}^{surface-force3,4}$ and $\dot{W}^{surface-force5,6}$ are expressed as:

$$\begin{aligned}\dot{W}^{surface-force3,4} &\approx \\ &\left[\sigma_{2j}(x_1, x_2+\Delta x_2/2, x_3)\right]\Delta x_3 \Delta x_1 V_j(x_1, x_2+\Delta x_2/2, x_3) \\ &-\left[\sigma_{2j}(x_1, x_2-\Delta x_2/2, x_3)\right]\Delta x_3 \Delta x_1 V_j(x_1, x_2-\Delta x_2/2, x_3),\end{aligned} \tag{5.69}$$

$$\begin{aligned}\dot{W}^{surface-force5,6} &\approx \\ &\left[\sigma_{3j}(x_1, x_2, x_3+\Delta x_3/2)\right]\Delta x_1 \Delta x_2 V_j(x_1, x_2, x_3+\Delta x_3/2) \\ &-\left[\sigma_{3j}(x_1, x_2, x_3-\Delta x_3/2)\right]\Delta x_1 \Delta x_2 V_j(x_1, x_2, x_3-\Delta x_3/2).\end{aligned} \tag{5.70}$$

To simplify the expression of (5.68)–(5.70), we express the components of the stress tensor and those of the velocity components at the face centers using Taylor series expansion up to first order, with the centroid of the control volume as the base location. It can be verified that the expressions in (5.68), (5.69), and (5.70) simplify to

$$\begin{aligned}\dot{W}^{surface-force1,2} &\approx \frac{\partial\left(\sigma_{1k} V_k\right)}{\partial x_1}\Delta x_1 \Delta x_2 \Delta x_3, \\ \dot{W}^{surface-force3,4} &\approx \frac{\partial\left(\sigma_{2k} V_k\right)}{\partial x_2}\Delta x_1 \Delta x_2 \Delta x_3, \\ \dot{W}^{surface-force5,6} &\approx \frac{\partial\left(\sigma_{3k} V_k\right)}{\partial x_3}\Delta x_1 \Delta x_2 \Delta x_3,\end{aligned} \tag{5.71}$$

where all derivatives are at the centroid of the fluid element $A'B'C'D'E'F'G'H'$ (x_1, x_2, x_3). Using (5.71) in (5.67), the net rate at which all external surface forces do work on the fluid element $A'B'C'D'E'F'G'H'$ is expressed as:

$$\dot{W}^{surfaceforce} \approx \left[\frac{\partial\left(\sigma_{1k} V_k\right)}{\partial x_1}+\frac{\partial\left(\sigma_{2k} V_k\right)}{\partial x_2}+\frac{\partial\left(\sigma_{3k} V_k\right)}{\partial x_3}\right]\Delta x_1 \Delta x_2 \Delta x_3$$

$$\approx \frac{\partial\left(\sigma_{mk} V_k\right)}{\partial x_m} \Delta x_1 \Delta x_2 \Delta x_3. \tag{5.72}$$

Using (5.72) and the expression for $\dot{W}^{bodyforce}$ from (5.66) in (5.65) we arrive at the expression for $\dot{W}$:

$$\dot{W} \approx \left[\rho g_k V_k + \frac{\partial\left(\sigma_{mk} V_k\right)}{\partial x_m}\right] \Delta x_1 \Delta x_2 \Delta x_3. \tag{5.73}$$

We now turn our attention to the quantity $\dot{Q}$ (5.64), the rate at which heat is being added to the fluid element $A'B'C'D'E'F'G'H'$. $\dot{Q}$ collectively represents the net rate of heat addition to the fluid element due to the process of (i) volumetric heating and (ii) conduction.

$$\dot{Q} = \dot{Q}^{conduction} + \dot{Q}^{volumetricheating}. \tag{5.74}$$

The term *volumetric heating* includes all those processes which generate and add heat to the fluid particles within the volume of the fluid element. Electrical heating and radiative heating are some examples of volumetric heating. The general mathematical expression of volumetric heating is

$$\dot{Q}^{volumetricheating} \approx \rho(x_1, x_2, x_3) \Delta x_1 \Delta x_2 \Delta x_3 \dot{S}(x_1, x_2, x_3), \tag{5.75}$$

where $\dot{S}(x_1, x_2, x_3)$ represents the rate of volumetric heat addition per unit mass at the centroid of the fluid element. Its specific mathematical form would depend on the exact physical process of heat generation.

In contrast to $\dot{Q}^{volumetricheating}$, $\dot{Q}^{conduction}$ represents the net rate at which heat is being added to the fluid element A'B'C'D'E'F'G'H' due to the heat exchange process happening between this fluid element and its surroundings. This exchange happens through the six external surfaces of the fluid element A'B'C'D'E'F'G'H'.

$$\begin{aligned} \dot{Q}^{conduction} &= \dot{Q}^{conduction,1} + \dot{Q}^{conduction,2} \\ &+ \dot{Q}^{conduction,3} + \dot{Q}^{conduction,4} \\ &+ \dot{Q}^{conduction,5} + \dot{Q}^{conduction,6}, \end{aligned} \tag{5.76}$$

where each of the terms on the right-hand side of (5.76) represents the rate at which heat is *added to* the fluid element through the respective face. To represent this process mathematically, we introduce a new vector quantity called the *heat flux per unit area*. We denote this quantity by the symbol $\underline{q}(\underline{X}, t)$. For our analysis and derivations, it is deemed as an Eulerian field. Consider a small area ΔA centered at location $(\underline{X})$ with its outward normal unit vector being $\hat{e}_n$ (Fig. 5.12). The quantity $\underline{q} \cdot \hat{e}_n \Delta A$ represents the heat flowing through the small area per unit area in the direction of $\hat{e}_n$ due to the process of conduction. The SI units of $\underline{q}$ are $J s^{-1} m^{-2}$.

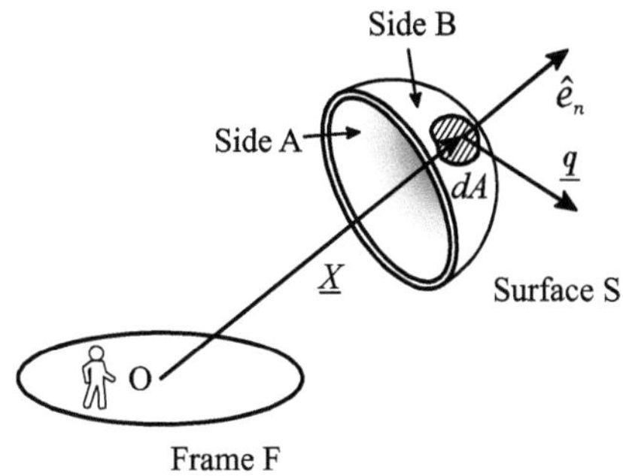

Fig. 5.12 The heat flux per unit area vector: $\underline{q}$

Since the face areas of the fluid element under consideration are all small, we approximate the rate at which heat is *leaving* the fluid element through a particular face by taking the product of the face area and the face-normal component of the heat flux per unit area vector at the centroid of the face. Thus,

$$\begin{aligned}
\dot{Q}^{conduction,1} &\approx -\Big[\underline{q}(x_1 + \Delta x_1/2, x_2, x_3)\Big] \cdot (\hat{e}_1)\Delta x_2 \Delta x_3, \\
&\approx -q_1(x_1 + \Delta x_1/2, x_2, x_3)\Delta x_2 \Delta x_3, \\
\dot{Q}^{conduction,2} &\approx \Big[\underline{q}(x_1 - \Delta x_1/2, x_2, x_3)\Big] \cdot (\hat{e}_1)\Delta x_2 \Delta x_3, \\
&\approx q_1(x_1 - \Delta x_1/2, x_2, x_3)\Delta x_2 \Delta x_3, \\
\dot{Q}^{conduction,3} &\approx -\Big[\underline{q}(x_1, x_2 + \Delta x_2/2, x_3)\Big] \cdot (\hat{e}_2)\Delta x_3 \Delta x_1, \\
&\approx -q_2(x_1, x_2 + \Delta x_2/2, x_3)\Delta x_3 \Delta x_1, \\
\dot{Q}^{conduction,4} &\approx \Big[\underline{q}(x_1, x_2 - \Delta x_2/2, x_3)\Big] \cdot (\hat{e}_2)\Delta x_3 \Delta x_1, \\
&\approx q_2(x_1, x_2 - \Delta x_2/2, x_3)\Delta x_3 \Delta x_1, \\
\dot{Q}^{conduction,5} &\approx -\Big[\underline{q}(x_1, x_2, x_3 + \Delta x_3/2)\Big] \cdot (\hat{e}_3)\Delta x_1 \Delta x_2, \\
&\approx -q_3(x_1, x_2, x_3 + \Delta x_3/2)\Delta x_1 \Delta x_2, \\
\dot{Q}^{conduction,6} &\approx \Big[\underline{q}(x_1, x_2, x_3 - \Delta x_3/2)\Big] \cdot (\hat{e}_3)\Delta x_1 \Delta x_2, \\
&\approx q_3(x_1, x_2, x_3 - \Delta x_3/2)\Delta x_1 \Delta x_2,
\end{aligned} \tag{5.77}$$

where q_1, q_2 and q_3 represent the three scalar components of the local $\underline{q}$ vector. We express the components of the heat flux vector at the face centers using appropriate Taylor series expansions up to first order, with the centroid (x_1, x_2, x_3) of the fluid element as the base location. Accordingly:

$$\begin{aligned}
q_1(x_1 + \Delta x_1/2, x_2, x_3) &\approx q_1(x_1, x_2, x_3) + \frac{\partial q_1}{\partial x_1}\frac{\Delta x_1}{2}, \\
q_1(x_1 - \Delta x_1/2, x_2, x_3) &\approx q_1(x_1, x_2, x_3) - \frac{\partial q_1}{\partial x_1}\frac{\Delta x_1}{2},
\end{aligned}$$

$$
\begin{aligned}
q_2(x_1, x_2 + \Delta x_2/2, x_3) &\approx q_2(x_1, x_2, x_3) + \frac{\partial q_2}{\partial x_2}\frac{\Delta x_2}{2},\\
q_2(x_1, x_2 - \Delta x_2/2, x_3) &\approx q_2(x_1, x_2, x_3) - \frac{\partial q_2}{\partial x_2}\frac{\Delta x_2}{2},\\
q_3(x_1, x_2, x_3 + \Delta x_3/2) &\approx q_3(x_1, x_2, x_3) + \frac{\partial q_3}{\partial x_3}\frac{\Delta x_3}{2},\\
q_3(x_1, x_2, x_3 - \Delta x_3/2) &\approx q_3(x_1, x_2, x_3) - \frac{\partial q_3}{\partial x_3}\frac{\Delta x_3}{2},
\end{aligned}
\tag{5.78}
$$

where all derivatives of various scalar components of $\underline{q}$ are at location (x_1, x_2, x_3). Using the approximations of (5.78), we arrive at:

$$
\begin{aligned}
\dot{Q}^{conduction,1} + \dot{Q}^{conduction,2} &\approx -\frac{\partial q_1}{\partial x_1}\Delta x_1 \Delta x_2 \Delta x_3,\\
\dot{Q}^{conduction,3} + \dot{Q}^{conduction,4} &\approx -\frac{\partial q_2}{\partial x_2}\Delta x_1 \Delta x_2 \Delta x_3,\\
\dot{Q}^{conduction,5} + \dot{Q}^{conduction,6} &\approx -\frac{\partial q_3}{\partial x_3}\Delta x_1 \Delta x_2 \Delta x_3.
\end{aligned}
\tag{5.79}
$$

Using (5.79) in (5.76) we arrive at the following expression of $\dot{Q}^{conduction}$.

$$
\begin{aligned}
\dot{Q}^{conduction} &\approx -\left[\frac{\partial q_1}{\partial x_1} + \frac{\partial q_2}{\partial x_2} + \frac{\partial q_3}{\partial x_3}\right]\Delta x_1 \Delta x_2 \Delta x_3\\
&\approx -\frac{\partial q_k}{\partial x_k}\Delta x_1 \Delta x_2 \Delta x_3.
\end{aligned}
\tag{5.80}
$$

Using (5.80) and (5.75) in (5.74) we arrive at the mathematical form of $\dot{Q}$:

$$
\dot{Q} \approx \rho \dot{S} \Delta x_1 \Delta x_2 \Delta x_3 - \frac{\partial q_k}{\partial x_k}\Delta x_1 \Delta x_2 \Delta x_3,
\tag{5.81}
$$

where ρ and $\dot{S}$ are computed at the centroid of the fluid element. With the expressions of $\dot{Q}$ (5.81) and $\dot{W}$ (5.73) now available, (5.64) is expressed as:

$$
\begin{aligned}
\frac{\partial \left[\rho \left(e + \frac{1}{2}V_k V_k\right)\right]}{\partial t} + \frac{\partial \left\{\left[\rho \left(e + \frac{1}{2}V_k V_k\right)\right] V_j\right\}}{\partial x_j} &= \rho g_k V_k + \frac{\partial \left(\sigma_{mk} V_k\right)}{\partial x_m}\\
&+ \rho \dot{S} - \frac{\partial q_k}{\partial x_k}.
\end{aligned}
\tag{5.82}
$$

Further, taking the product of $\left(e + \frac{1}{2}V_k V_k\right)$ with the governing equation of density (5.29) and subtracting the resulting equation from (5.82) results into the following form of the governing equation of $(e + \frac{1}{2}V_k V_k)$:

$$\frac{\partial\left(e+\frac{1}{2}V_kV_k\right)}{\partial t}+V_m\frac{\partial\left(e+\frac{1}{2}V_kV_k\right)}{\partial x_m}=g_kV_k+\frac{1}{\rho}\frac{\partial\left(\sigma_{mk}V_k\right)}{\partial x_m}+\dot{S}-\frac{1}{\rho}\frac{\partial q_k}{\partial x_k}. \tag{5.83}$$

Equation (5.83) represents the transport equation of total energy per unit mass $\left(e+\frac{1}{2}V_kV_k\right)$. We now wish to use (5.83) to derive the transport equation of e, itself. This can be achieved by first deriving the transport equation of kinetic energy per unit mass ($\frac{V_kV_k}{2}$) and then subtracting it from the transport equation of total energy per unit mass (5.83).

The transport equation of $\frac{V_kV_k}{2}$ is obtained by first taking the product of the transport equation of V_1 (5.63) and V_1, taking the product of the transport equation of V_2 and V_2, and taking the product of the transport equation of V_3 and V_3

$$\begin{aligned}
V_1\rho\frac{\partial V_1}{\partial t}+V_1\rho V_k\frac{\partial V_1}{\partial x_k}&=V_1\rho g_1+V_1\frac{\partial\sigma_{k1}}{\partial x_k},\\
V_2\rho\frac{\partial V_2}{\partial t}+V_2\rho V_k\frac{\partial V_2}{\partial x_k}&=V_2\rho g_2+V_2\frac{\partial\sigma_{k2}}{\partial x_k},\\
V_3\rho\frac{\partial V_3}{\partial t}+V_3\rho V_k\frac{\partial V_3}{\partial x_k}&=V_3\rho g_3+V_3\frac{\partial\sigma_{k3}}{\partial x_k}.
\end{aligned} \tag{5.84}$$

and the adding the three resulting equations to arrive at

$$\left(\frac{\partial}{\partial t}+V_m\frac{\partial}{\partial x_m}\right)\left(\frac{V_kV_k}{2}\right)=V_mg_m+\frac{1}{\rho}V_m\frac{\partial\sigma_{km}}{\partial x_k}. \tag{5.85}$$

Equation (5.85) is the transport equation of kinetic energy per unit mass. Subtracting (5.85) from (5.83), we arrive at the transport equation of e:

$$\frac{\partial e}{\partial t}+V_k\frac{\partial e}{\partial x_k}=\frac{1}{\rho}\sigma_{km}\frac{\partial V_m}{\partial x_k}+\dot{S}-\frac{1}{\rho}\frac{\partial q_k}{\partial x_k}, \tag{5.86}$$

which can be expressed using the material derivative operator as:

$$\frac{De}{Dt}=\frac{1}{\rho}\sigma_{km}\frac{\partial V_m}{\partial x_k}+\dot{S}-\frac{1}{\rho}\frac{\partial q_k}{\partial x_k}. \tag{5.87}$$

The left-hand side of the equation represents the rate of change of internal energy per unit mass following a local fluid particle. The right-hand side identifies three processes due to which e changes. The first process on the rhs represents the rate at which the surface forces do work. The second term is the rate of heat addition due to volumetric heating. The third process represents the rate of heat addition to the fluid particle from the surroundings. If we restrict ourselves to fluids which have their specific heat of constant volume (C_v) a constant, internal energy can be expressed as

$$e=C_vT. \tag{5.88}$$

For such fluids, the transport equation of e can be expressed alternatively as the transport equation of T, itself:

$$\frac{DT}{Dt} = \frac{1}{\rho C_v}\sigma_{km}\frac{\partial V_m}{\partial x_k} + \frac{\dot{S}}{C_v} - \frac{1}{\rho C_v}\frac{\partial q_k}{\partial x_k}. \tag{5.89}$$

5.7 Closure of Governing Equations

In Chap. 3, we identified six scalar-dependent variables with which we describe the fluid motion. These variables are: the velocity vector ($\underline{V}$), density (ρ), pressure (p), and temperature(T). In this chapter, so far, we have derived the governing equations of ρ, $\underline{V}$, and T. These equations are (5.30), (5.63), and (5.89) and are summarized here for further reference.

$$\begin{aligned}
\frac{D\rho}{Dt} &= -\rho\frac{\partial V_k}{\partial x_k},\\
\frac{DV_i}{Dt} &= g_i + \frac{1}{\rho}\frac{\partial \sigma_{ji}}{\partial x_j},\\
\frac{DT}{Dt} &= \frac{1}{\rho C_v}\sigma_{mk}\frac{\partial V_k}{\partial x_m} + \frac{\dot{S}}{C_v} - \frac{1}{\rho C_v}\frac{\partial q_k}{\partial x_k}.
\end{aligned} \tag{5.90}$$

Equation (5.90) is a set of five scalar equations. During the process of deriving these equations, however, we had to introduce two new tensors: the stress tensor $\underline{\sigma}$ and the heat flux per unit area vector $\underline{q}$. The stress tensor, being a second-order tensor, introduces, in general, nine new scalar Eulerian variables. The heat flux per unit area vector introduces, in general, three new scalar Eulerian variables. We call these new variables as the *secondary* variables that we require to have the complete description of the flow field (in contrast, we call $\underline{V}$, p, T, and ρ as the *primary* variables). Thus, the equation set (5.90) involves, in total, 18 unknowns. With the number of unknown scalars exceeding the number of scalar equations, (5.90) is *mathematically unclosed*. To make this equation set mathematically closed and thus viable for any further theoretical or computational analysis, we must have additional new equations relating the secondary variables to the primary variables.

Based on the nature of different origins of the surface forces, the stress tensor at a point in a fluid medium is expressed as a sum of two parts. The first part, which is called the *pressure stress tensor*, arises because of the existence of pressure in a fluid medium. The other part of the stress tensor arises because of the viscous nature of the fluid medium. The tensor equation relating the two parts to the stress tensor $\underline{\sigma}$ is:

$$\underline{\sigma} = -p\underline{I} + \underline{\tau}, \tag{5.91}$$

where $\underline{I}$ represents the identity tensor of order two. In the Cartesian coordinate system, $\underline{I}$ is expressed as

$$\underline{I} = \hat{e}_1\hat{e}_1 + \hat{e}_2\hat{e}_2 + \hat{e}_3\hat{e}_3. \tag{5.92}$$

The symbol $\underline{\tau}$ represents the part of the stress tensor arising because of the viscous nature of the fluid medium. This tensor is called the *viscous stress tensor*. In a Cartesian coordinate system, the (ij)th component of the stress tensor is expressed using the Kronecker delta symbol:

$$\sigma_{ij} = -p\delta_{ij} + \tau_{ij}. \tag{5.93}$$

Even though p is a positive definite quantity ($p \geq 0$), the normal stress components arising due to pressure are always negative, implying that the corresponding normal forces are always acting toward the surface under consideration. The pressure stress tensor ($-p\underline{I}$) is an *isotropic* second-order tensor (all shear components zero and all the normal components being identical). On the other hand, in general, all scalar components (normal or shear) of the viscous stress tensor may be non-zero.

From the point of view of mathematical closure, the pressure stress tensor does not add any new variables (compared to the original set of Eulerian variables: p ρ, T, $\underline{V}$. The viscous stress tensor in (5.90) is still an unknown tensor. Even though this tensor, in general, can introduce nine secondary unknowns, it can be shown that in fluids having no body couples, this tensor must be symmetric. Thus, it introduces only six new secondary unknowns (more on this is Sect. 5.7.1).

There are no fundamental laws based on which we can completely ascertain how the viscous stress tensor depends on the primary Eulerian variables. Thus, we must rely on experimental observations and measurements to find approximate relationships between the viscous stress tensor and the primary flow variables. Further, these relationships (which are also called the *constitutive relationships*) are not found to be universal in nature. These may differ from one fluid to another. In this book, we focus on a special class of fluids called the *Newtonian fluids*. In Sect. 5.7.2, we present the constitutive relationship for the viscous stress tensor applicable to a Newtonian fluid.

5.7.1 Symmetry of the Stress Tensor

So far, we have applied the (i) law of conservation of mass, (ii) Euler's first axiom, and (iii) the first law of thermodynamics to the small fluid element $A'B'C'D'E'F'G'H'$ (Fig. 5.3). From the point of view of the primary variables, these three laws/axioms were useful, as they led to the governing equations of the primary variables (ρ, $\underline{V}$, and T). Furthermore, the rotational motion of the small fluid element $A'B'C'D'E'F'G'H'$ can be described using Euler's second axiom of motion, which states that *the rate of change of angular momentum of a body with respect to an inertial reference frame and about its center of mass equals the net moment of all the external forces acting on the body about its center of mass*. However, all those external forces which have their lines of action passing through the center of mass of the fluid element (same as the centroid of the fluid element in Fig. 5.3) do not contribute to this net moment. Accordingly, the resultant gravitational force on the fluid element, which acts at the centroid of the fluid element, is inconsequential. Similarly, the surface force vectors arising because of the normal stresses acting at the centroids of the six faces of the fluid element (Figs. 5.9, 5.10, 5.11) will also be inconsequential, because their lines of action, too, pass through the center of mass of the fluid element. However, surface forces arising due to the shear stress components acting on the six faces do contribute to the net moment vector, and those relevant stress components would indeed appear in the equation resulting from Euler's second axiom being applied to the fluid element.

Theoretically, it is possible that the resultant of the body forces on a small fluid element can be such that the equivalent force system is comprised of not only a resultant force acting at the center of the fluid element (which has zero moment about the center of mass) but also a net non-zero moment due to *body couples*. If such a moment due to body couples exists, it must be included while applying Euler's second axiom on the fluid element under consideration. However, the common fluids of engineering interest (air and water) are known to be free of such moments of body couples. For such fluids, application of Euler's second axiom leads to the conclusion that the stress tensor ($\underline{\sigma}$) must be a symmetric tensor (detailed algebra not included here, full derivation available in [2]).

$$\underline{\sigma} = \underline{\sigma}^T. \tag{5.94}$$

Since the pressure stress tensor ($-p\underline{I}$) itself is isotropic, the symmetry of $\underline{\sigma}$ implies that the viscous stress tensor ($\underline{\tau}$) must be symmetric. In the rest of this book, we restrict ourselves only to those fluids that do not have body couples and consequently, the stress tensor is always symmetric. For such fluids, the viscous stress tensor introduces only six secondary variables in (5.90).

5.7.2 A Closure Equation for the Stress Tensor

In this book, now onward, we focus entirely on a class of fluids called the *Newtonian fluids* in which the stress tensor follows the following constitutive relationship for the viscous stress tensor

$$\underline{\tau} = 2\mu\underline{S} + \lambda trace(\underline{S})\underline{I} = 2\mu\underline{S} + \lambda\left(\underline{\nabla} \cdot \underline{V}\right)\underline{I}. \tag{5.95}$$

where $\underline{S}$ is the strain-rate tensor (4.41) and trace($\underline{S}$) means $S_{ii}(= S_{11} + S_{22} + S_{33})$, which also equals the divergence of the velocity vector ($\underline{\nabla} \cdot \underline{V}$). The symbols μ and λ are called the *first* and the *second coefficients of viscosity*. Since both the strain-rate tensor and the identity tensor ($\underline{I}$) are symmetric tensors, the rhs of (5.95) is always symmetric. This ensures that $\underline{\tau}$ for Newtonian fluids is always symmetric. Air and water are the most common Newtonian fluids.

It is a common practice to approximate $\lambda = -\frac{2}{3}\mu$. Thus, (5.95) simplifies to

$$\underline{\tau} = 2\mu \underline{S} - \frac{2}{3}\mu \left(\underline{\nabla} \cdot \underline{V}\right) \underline{I}. \tag{5.96}$$

There are many engineering applications in which the fluid density does not change significantly either with time or space (constant-density flow fields). For such flow fields, the lhs (5.30) is negligible, implying that $\underline{\nabla} \cdot \underline{V} \approx 0$. For such flow fields, the Newtonian constitutive relationship (5.96) simplifies further to

$$\underline{\tau} = 2\mu \underline{S}. \tag{5.97}$$

In such flow fields, the actual value of λ is inconsequential, and μ is the only coefficient of viscosity that matters.

The numerical value of μ is measured experimentally for different fluids. For both liquids and gases, it is known to have a dependence on temperature. For liquids, it tends to decrease with increasing temperature. For gases, it tends to increase with increasing temperature. For flow fields with no significant variation in the temperature field, μ is treated as a known constant. The values of μ for water and air at 298 K are 8.92110^{-4} kgm^{-1}s^{-1} and 1.83610^{-5} kgm^{-1}s^{-1}, respectively. A fluid with a higher value of μ is said to be more *viscous*. The first coefficient of viscosity is also called the *coefficient of dynamic viscosity* of the fluid.

Equation (5.97) shows that an individual normal viscous stress component (τ_{11}, τ_{22} or τ_{33}) depends on the corresponding normal strain-rate component, as well as, on the local dilatation-rate (S_{kk}).

$$\begin{aligned} \tau_{11} &= 2\mu S_{11} - \frac{2\mu}{3} S_{kk}, \\ \tau_{22} &= 2\mu S_{22} - \frac{2\mu}{3} S_{kk}, \\ \tau_{33} &= 2\mu S_{33} - \frac{2\mu}{3} S_{kk}. \end{aligned} \tag{5.98}$$

In contrast, an individual shear stress component (τ_{12}, τ_{23} or τ_{31}) of the viscous stress tensor depends on the corresponding shear strain-rate component alone.

$$\tau_{12} = 2\mu S_{12}, \; \tau_{23} = 2\mu S_{23}, \; \tau_{31} = 2\mu S_{31}. \tag{5.99}$$

Equation (5.99) shows that for Newtonian fluids, the slope of the curve representing variation of a shear stress component to the corresponding component of the local strain-rate tensor is twice the coefficient of dynamic viscosity of the fluid.

$$\frac{\sigma_{12}}{S_{12}} = \frac{\sigma_{23}}{S_{23}} = \frac{\sigma_{31}}{S_{31}} = 2\mu. \tag{5.100}$$

For a chosen Newtonian fluid at a constant temperature, this slope (2μ) is a constant. In contrast, there do exist several other known fluids for which the slope of the curve representing the variation of a component of the shear stress tensor versus the corresponding component of the shear strain-rate tensor is not a constant (*non-Newtonian* fluids) even at constant temperature.

5.7.3 A Closure Equation for the Heat Flux Vector

For closure of the heat flux per unit area vector ($\underline{q}$), the following constitutive equation is used:

$$\underline{q} = -k\underline{\nabla}T, \tag{5.101}$$

where T is the local fluid temperature, and k is the thermal conductivity of the fluid. This relationship is commonly known as the *Fourier's law of heat conduction*. The negative sign present on the right-hand side (rhs) of (5.101) implies that the direction of heat transfer is opposite to the gradient of the temperature field. In other words, the process of conduction causes heat to flow from a higher-temperature region to lower-temperature regions.

5.8 The Navier–Stokes Equation Set

Using (5.91) to split the stress tensor into the pressure and the viscous stress tensors, the set of governing equations of fluid motion (5.90) for the Newtonian fluids is expressed as:

$$\begin{aligned}
\frac{\partial \rho}{\partial t} + V_k \frac{\partial \rho}{\partial x_k} &= -\rho \frac{\partial V_k}{\partial x_k}, \\
\frac{\partial V_i}{\partial t} + V_k \frac{\partial V_i}{\partial x_k} &= -\frac{1}{\rho}\frac{\partial p}{\partial x_i} + g_i + \frac{1}{\rho}\frac{\partial \tau_{ji}}{\partial x_j}, \\
\frac{\partial T}{\partial t} + V_m \frac{\partial T}{\partial x_m} &= -\frac{p}{\rho C_v}\frac{\partial V_k}{\partial x_k} + \frac{1}{\rho C_v}\tau_{mk}\frac{\partial V_k}{\partial x_m} + \frac{\dot{S}}{C_v} \\
&\quad -\frac{1}{\rho C_v}\frac{\partial q_k}{\partial x_k},
\end{aligned} \tag{5.102}$$

where, $\tau_{ji} = 2\mu s_{ji} - \frac{2\mu}{3}\delta_{ji}\frac{\partial V_k}{\partial x_k}$ (5.95) and $q_i = -k\frac{\partial T}{\partial x_i}$ (5.101). In the governing equation of V_i, the decomposition of the stress tensor in terms of the pressure stress and viscous shear stress has led to a term involving the gradient of pressure and another one involving the gradient of the viscous stress tensor. These terms represent the net pressure force and net viscous force per unit mass of fluid. In the governing equation of T, the same decomposition leads to the appearance of terms $-p\frac{\partial V_k}{\partial x_k}$ and

$\tau_{mk}\frac{\partial V_k}{\partial x_m}$. The first term represents the rate of work done per unit mass by the pressure forces and is called the *pressure-dilatation* process. On the other hand, $\tau_{mk}\frac{\partial V_k}{\partial x_m}$ represents the rate at which viscous forces do work (per unit mass). This term is called the *viscous heating* process. Since the viscous stress tensor is symmetric, it can be shown (the tensor identity (1.76) would prove useful toward this demonstration) that $\tau_{mk}\frac{\partial V_k}{\partial x_m} = \tau_{mk}S_{km}$, where S_{km} is the (km)th component of the local strain-rate tensor.

$$\begin{aligned}\tau_{mk}\frac{\partial V_k}{\partial x_m} &= \tau_{mk}\left(S_{mk} + W_{mk}\right) = \tau_{mk}S_{mk} + \tau_{mk}W_{mk} = \tau_{mk}S_{mk} \\ &= \underline{\tau} : \underline{S}^T = \underline{\tau} : \underline{S}.\end{aligned} \tag{5.103}$$

where W_{mk} is the (km)th component of the rotation-rate tensor.

Equation set (5.102) is a now set of five scalar partial differential equations in six unknowns V_1, V_2, V_3, p, T, and ρ. For a general compressible medium, the sixth equation must be the state equation relating pressure to a function of temperature and density

$$p = f(\rho, T). \tag{5.104}$$

If the medium is a perfect gas, then the state equation is:

$$p = \rho RT, \tag{5.105}$$

where R represents gas constant $(Jkg^{-1}K^{-1})$. Adding (5.105) to the set (5.102) now closes the system of six equations in as many unknowns. This closed set is called the *Navier–Stokes* equation set. Except for the state equation, which is a non-linear algebraic equation, the other equations are non-linear partial differential equations. Further, all equations of the Navier–Stokes equation set are coupled (each equation involves more than one unknown variable). Even though the Navier–Stokes equation set is mathematically closed, there is no general closed-form solution known for this equation set.

Using the index notation discussed in Chap. 1, it can be shown that the Navier–Stokes equation set can be expressed in the following form, which is independent of the choice of the working coordinate system:

$$\begin{aligned}\frac{\partial \rho}{\partial t} + \left(\underline{V}\cdot\underline{\nabla}\right)\rho &= -\,\rho\left(\underline{\nabla}\cdot\underline{V}\right), \\ \rho\frac{\partial \underline{V}}{\partial t} + \rho\left(\underline{V}\cdot\underline{\nabla}\right)\underline{V} &= -\,\underline{\nabla}p + \rho\underline{g} + \underline{\nabla}\cdot\underline{\tau}, \\ \rho C_v\frac{\partial T}{\partial t} + \rho C_v\left(\underline{V}\cdot\underline{\nabla}\right)\underline{T} &= -\,p\left(\underline{\nabla}\cdot\underline{V}\right) + \underline{\tau} : \underline{S} + \rho\dot{S} + \underline{\nabla}\cdot\left(k\underline{\nabla}T\right),\end{aligned} \tag{5.106}$$

where $\underline{\tau} = 2\mu\underline{S} - \frac{2\mu}{3}\left(\underline{\nabla}\cdot\underline{V}\right)\underline{I}$.

5.8.1 Incompressible Navier–Stokes Equation Set

Flow in which $\frac{D\rho}{Dt} = 0$ at all locations and at all times are called incompressible flows. Thus, the definition of incompressibility requires the rate of change of density following each particle at all times be zero. Otherwise, the flow field is said to be compressible.

There are many flow fields in which both density and temperature variations are negligible in both time and space ($\frac{\partial \rho}{\partial t}, \frac{\partial T}{\partial t} \approx 0$ and $\frac{\partial \rho}{\partial x_i}, \frac{\partial T}{\partial x_i} \approx 0$). In such flow fields, the continuity equation simplifies to

$$0 \approx \underline{\nabla} \cdot \underline{V}. \tag{5.107}$$

In such flow fields, there is no need to solve the transport equation of T. Further, as a consequence of the simplified continuity equation (5.107), the constitutive equation for the viscous stress (5.96) simplifies to

$$\underline{\tau} \approx 2\mu \underline{S}. \tag{5.108}$$

It can be verified that (5.108) when used to substitute for the viscous stress tensor in the transport equation of $\underline{V}$ in (5.106) leads to the following simplified form of the equation:

$$\rho \frac{\partial \underline{V}}{\partial t} + \rho \left(\underline{V} \cdot \underline{\nabla}\right) \underline{V} = -\underline{\nabla} p + \rho \underline{g} + \mu \nabla^2 \underline{V}, \tag{5.109}$$

where μ is a known constant.

In summary, for flow fields with negligible density and temperature variations, the governing equation set is:

$$\begin{aligned} \underline{\nabla} \cdot \underline{V} &= 0, \\ \rho \frac{\partial \underline{V}}{\partial t} + \rho \left(\underline{V} \cdot \underline{\nabla}\right) \underline{V} &= -\underline{\nabla} p + \rho \underline{g} + \mu \nabla^2 \underline{V}. \end{aligned} \tag{5.110}$$

This is a set of 4 scalar equations with as many scalar unknowns ($\underline{V}$ and p). The equation set (5.110) is called the *incompressible Navier–Stokes equation* set. Even though the governing equation set (5.110) is still coupled, the number of equations and the number of unknowns are fewer compared to the equation set (5.106). Further, the non-linearity of the partial differential equations is restricted only to the transport equation of $\underline{V}$. The simplified continuity equation is linear. Despite these simplifications, still no general closed-form solution is known even for the incompressible Navier–Stokes equation set (5.110).

Equation set (5.110) is applicable for liquid flows which do not involve any significant heat transfer. The density variations in liquid flows are also negligible. The condition of no heat transfer ensures that the variations in temperature, too, are negligible. Equation set (5.110) is also applicable for a special category of gaseous flows. Such a gaseous flow can be identified by estimating the value of a non-dimensional quantity called the *Mach* number. The Mach number (M) is defined as the ratio of the characteristic timescale of the motion of the constituent molecules to the characteristic timescale of the motion of a typical fluid particle of the same flow field. This ratio, in turn, determines how fast pressure adapts in the flow field in response to the changes in the continuum velocity field. For gaseous flows having a small Mach number (typically, <0.3) and not involving any significant heat transfer, both density and temperature variations are found to be negligible. Then equation set (5.110) is suitable to describe such gaseous flow fields. The airflow past an automobile on a highway, that through the vent of a room air-conditioner or a hand-dryer, wind blowing past buildings are some typical scenarios wherein the flow of air is described by the incompressible Navier–Stokes set (5.110) up to acceptable accuracy, and we need not necessarily refer to the more general and the more elaborate form of the Navier–Stokes equation set (5.106).

In the rest of the book, we restrict ourselves to only such flow fields wherein the motion can be described using the incompressible Navier–Stokes equation set (5.110). Any further examination of flows with variable density and temperature is deemed outside the scope of this book. As we acknowledged earlier, we do not know a general closed-form solution for (5.110). However, with massive computational power made available by modern computers along with the simultaneous development of appropriate numerical techniques to solve even highly coupled partial differential equations, today, it is possible to obtain numerical solutions of the Navier–Stokes equations up to a desired level of accuracy for several types of flow fields. This branch of fluid mechanics is called *computational fluid dynamics* (CFD). Such numerical solutions can then be used to develop a better understanding of flow fields and to design engineering systems that interact with such flow fields. The study of such CFD techniques and their implementation, however, is outside the scope of this book. Instead, our focus in this book will now turn to some simple flow fields wherein the incompressible Navier–Stokes equation set itself can be further simplified to the extent that we can find an analytical solution. Even though they are applicable to only some special flow fields, examining those solutions will still provide opportunities to develop deeper insights into the mechanics of fluid flows.

Chapter 6
Interaction of a Flow Field with a Solid Surface

Many engineering applications involve the interaction of fluids with solid surfaces. The presence of an interacting solid surface influences the flow field. Specifically, this interaction imposes certain conditions on the velocity of those fluid particles which are in touch with the solid surface at a solid–fluid interface. Even though the solid surface imposes these velocity conditions on the interfacing fluid particles, these conditions (the boundary conditions) can influence the velocity as well as the pressure field over extended regions of the flow domain through the governing equations (the Navier–Stokes equation set). Further, it is also of interest to estimate the interaction forces generated between a solid and a fluid medium. The calculation of these forces would require a knowledge of the stress distribution along the fluid–solid interface. In this chapter, we first study the velocity conditions a solid surface imposes on fluid motion. Subsequently, we study how to find the interaction force between a fluid medium and a solid surface.

6.1 Velocity Constraints on a Flow Field on a Solid Surface

Let us consider an interface of a flow field and a solid object as shown in Fig. 6.1.

Let us focus our attention on a typical pair of particles, A and B, where A is a fluid particle and B is a solid particle. These particles are in touch with each other. We are observing the motion of these particles with respect to a frame F. Barring a few exceptions, most fluids are known to exhibit the following constraints,

$$\underline{V}_{A|F} = \underline{V}_{B|F}, \tag{6.1}$$

where the symbol $\underline{V}_{A|F}$ and $\underline{V}_{B|F}$ are the velocity vectors of the particles A and B with respect to the frame F. The vector equation (6.1) can be equivalently expressed as a combination of two equations. Taking the component of (6.1) along the local normal unit vector $\hat{e}_n$ (Fig. 6.1) results into the following scalar equation

© The Author(s), under exclusive license to Springer Nature Switzerland AG 2026
S. S. Sinha, *Fundamentals of Fluid Mechanics*,
https://doi.org/10.1007/978-3-032-04210-1_6

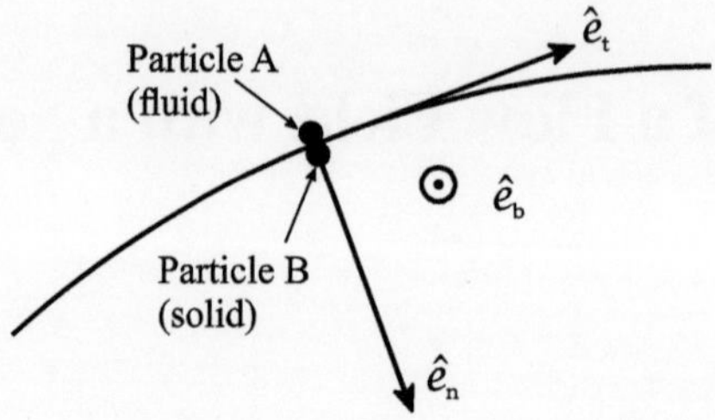

Fig. 6.1 Interaction of a solid and a fluid particle. The sizes of these particles have been exaggerated for illustration. $\hat{e}_n$ is the local surface-normal vector pointing toward the solid particle. $\hat{e}_t$ is the local surface-tangent, and $\hat{e}_b = \hat{e}_t \times \hat{e}_n$. A symbol of a circle with an embedded dot indicates a direction going into the plane of the paper

$$\underline{V}_{A|F} \cdot \hat{e}_n = \underline{V}_{B|F} \cdot \hat{e}_n. \tag{6.2}$$

Equation (6.2) is called as the *no-penetration* condition. Subtracting (6.2) from (6.1) results into

$$\underline{V}_{A|F} - \underline{V}_{A|F} \cdot \hat{e}_n = \underline{V}_{B|F} - \underline{V}_{B|F} \cdot \hat{e}_n. \tag{6.3}$$

Equation (6.3) is called as the *no-slip* condition existing at a solid–fluid interface. *N!no-slip* Even though (6.1), (6.2) and (6.3) are expressed here for a single pair of particles A and B, the no-slip and the no-penetration conditions must hold good for all such pairs of particles distributed over the entire solid–fluid interface shown in Fig. 6.1.

In the study of fluid mechanics, we define a special class of fluids called the *ideal* fluids, wherein the viscous stress tensor is zero, and the stress tensor arises entirely due to pressure.

$$\underline{\sigma} = -p\underline{I}. \tag{6.4}$$

In contrast, fluids which have a non-zero viscous stress tensor are called *real* fluids. Referring to the constitutive equation of the Newtonian fluids (5.96), an ideal fluid can alternatively be presented as a fluid with $\mu = 0$. Only the no-penetration condition is imposed with ideal fluids at a solid interface. In the absence of viscosity, the no-slip constraint does not exist.

6.2 Force System Exerted by Fluid on a Solid Surface

The force experienced by a solid surface due to the interacting fluid can be expressed using the distribution of the local stress tensor on the fluid side of a fluid–solid interface. Let us consider a pair of small but equal areas belonging to the fluid continuum and the interacting solid surface: dA_{fluid} and dA_{solid} (Fig. 6.2).

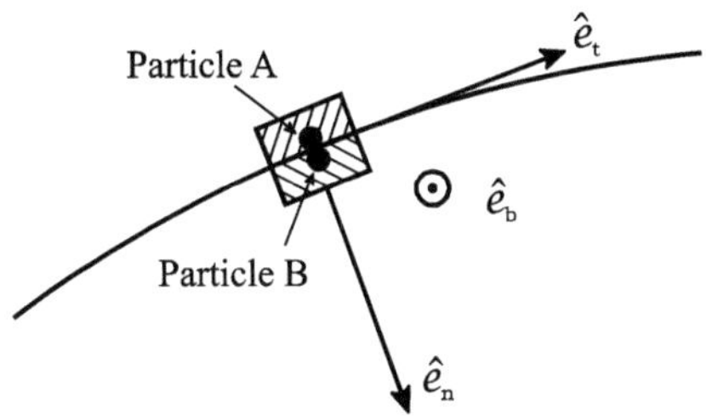

Fig. 6.2 Interaction of a small fluid element with a small solid element (shaded areas). The two elements interact through the faces dA_{fluid} (with outward unit normal $\hat{e}_n$) and dA_{solid} (with outward unit normal $-\hat{e}_n$). Particles A and B (same as shown in Fig. 6.1) are located at the centroids of dA_{fluid} and dA_{solid}, respectively

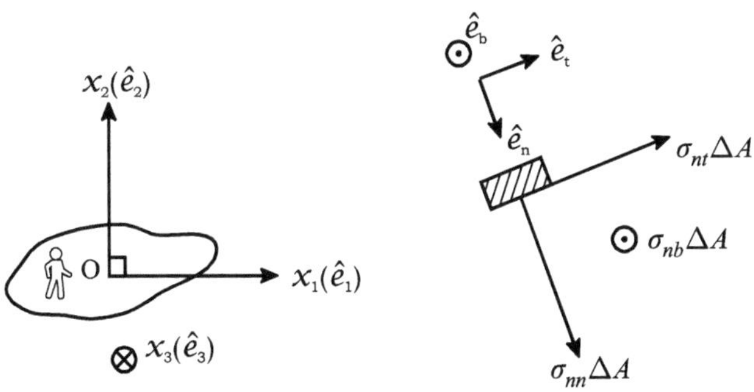

Fig. 6.3 External forces acting on the shaded fluid element through the area dA_{fluid} at the solid–fluid interface. A symbol of a circle with an embedded cross indicates a direction coming out of the plane of the paper

$$dA_{fluid} = dA_{fluid} = \Delta A. \tag{6.5}$$

In this figure, while $\hat{e}_n$ is the outward unit normal vector of dA_{fluid}, and $-\hat{e}_n$ is the outward unit normal vector of dA_{solid}. The locations of the particles A and B are the centroids of dA_{fluid} and dA_{solid}, respectively.

If $d\underline{F}$ represents the force exerted on dA_{fluid} by dA_{solid}, we wish to express $d\underline{F}$ in terms of the relevant fluid stress components existing in the fluid domain. In Fig. 6.2 we show three mutually perpendicular unit vectors $\hat{e}_n$ (local normal to the interface), $\hat{e}_t$ (the local surface-tangent unit vector) and $\hat{e}_b$ ($\hat{e}_b = \hat{e}_t \times \hat{e}_n$). Using these unit vectors, we express $d\underline{F}$ in terms of the relevant stress components as

$$d\underline{F} = \sigma_{nn} dA_{fluid}\hat{e}_n + \sigma_{nt} dA_{fluid}\hat{e}_t + \sigma_{nb} dA_{fluid}\hat{e}_b, \tag{6.6}$$

where a repeated index does not imply Einstein's summation (the symbol $\sigma_{nn} dA_{fluid}$ $\hat{e}_n$ represents just one term) (Fig. 6.3). Following Newton's third law, the local force exerted on dA_{solid} is $-d\underline{F}$. Thus, the net force acting on the solid surface ($\underline{F}^{solid}$) by the fluid can be computed and expressed in the frame-fixed Cartesian coordinate

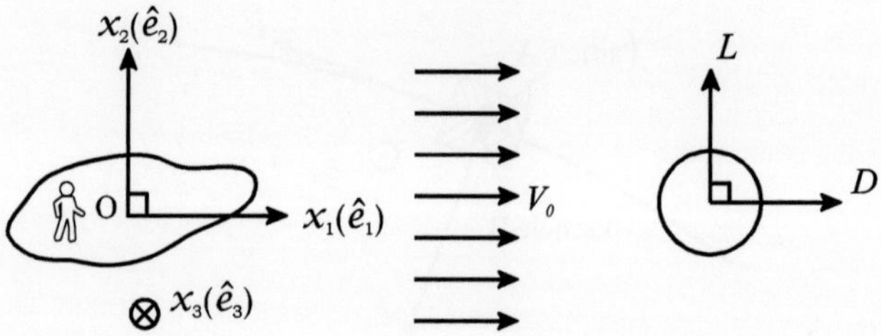

Fig. 6.4 Lift (L) and drag (D) forces acting on a solid object with uniform velocity in the far-upstream locations

system $Ox_1(\hat{e}_1)x_2(\hat{e}_2)x_3(\hat{e}_3)$ as:

$$\begin{aligned}\underline{F}^{solid} = &-\left[\iint_{A_{solid}} \left(\sigma_{nn}\hat{e}_n \cdot \hat{e}_1 + \sigma_{nt}\hat{e}_t \cdot \hat{e}_1 + \sigma_{nb}\hat{e}_b \cdot \hat{e}_1\right) dA_{solid}\right]\hat{e}_1 \\ &-\left[\iint_{A_{solid}} \left(\sigma_{nn}\hat{e}_n \cdot \hat{e}_2 + \sigma_{nt}\hat{e}_t \cdot \hat{e}_2 + \sigma_{nb}\hat{e}_b \cdot \hat{e}_2\right) dA_{solid}\right]\hat{e}_2 \\ &-\left[\iint_{A_{solid}} \left(\sigma_{nn}\hat{e}_n \cdot \hat{e}_3 + \sigma_{nt}\hat{e}_t \cdot \hat{e}_3 + \sigma_{nb}\hat{e}_b \cdot \hat{e}_3\right) dA_{solid}\right]\hat{e}_3, \end{aligned} \tag{6.7}$$

where we have used the relationship $dA_{fluid} = dA_{solid}$, and the integral is performed over the entire solid surface (A_{solid}). For a constant-density Newtonian fluid, using (5.98), the individual stress components are expressed

$$\sigma_{nn} = -p + 2\mu S_{nn} = -p + 2\mu\left[\frac{1}{2}\left(\frac{\partial V_n}{\partial x_n} + \frac{\partial V_n}{\partial x_n}\right)\right] = -p + 2\mu\frac{\partial V_n}{\partial x_n},$$

$$\sigma_{nt} = 2\mu\left[\frac{1}{2}\left(\frac{\partial V_t}{\partial x_n} + \frac{\partial V_n}{\partial x_t}\right)\right], \sigma_{nb} = 2\mu\left[\frac{1}{2}\left(\frac{\partial V_b}{\partial x_n} + \frac{\partial V_n}{\partial x_b}\right)\right], \tag{6.8}$$

where, again, a repeated index does not imply summation.

Many engineering applications involve a solid object being kept stationary (with respect to an inertial reference frame in context) in an otherwise uniform flow (Fig. 6.4). If the working coordinate system is oriented in a manner such that in the far-upstream condition, the velocity field is $V_o\hat{e}_1$, then the component of $\underline{F}^{solid}$ along $\hat{e}_1$ is called the *drag force* (D) experienced by the solid body. The component of $\underline{F}^{solid}$ perpendicular to $\hat{e}_1$ is called the *lift force* (L) experienced by the solid body. Further, the part of the drag force which arises due to the pressure stress is called the *pressure drag*,, and the part which arises due to the viscous stress is called the *viscous drag*.

$$D = \underline{F}^{solid} \cdot \hat{e}_1 = D_{pressure} + D_{viscous}, \tag{6.9}$$

where

$$\begin{aligned} D_{pressure} &= -\iint_{A_{solid}} \left(-p\hat{e}_n \cdot \hat{e}_1\right) dA_{solid} \\ &= \iint_{A_{solid}} \left(p\hat{e}_n \cdot \hat{e}_1\right) dA_{solid}. \end{aligned} \tag{6.10}$$

$$\begin{aligned} D_{viscous} = &-\iint_{A_{solid}} 2\mu \left(\frac{\partial V_n}{\partial x_n}\right) \hat{e}_n \cdot \hat{e}_1 dA_{solid} \\ &-\iint_{A_{solid}} \mu \left(\frac{\partial V_t}{\partial x_n} + \frac{\partial V_n}{\partial x_t}\right) \hat{e}_t \cdot \hat{e}_1 dA_{solid} \\ &-\iint_{A_{solid}} \mu \left(\frac{\partial V_b}{\partial x_n} + \frac{\partial V_n}{\partial x_b}\right) \hat{e}_b \cdot \hat{e}_1 dA_{solid}. \end{aligned} \tag{6.11}$$

Chapter 7
Hydrostatics

Hydrostatics is the study of mechanics of fluids at rest with respect to an inertial reference frame. Even though the fluid is at rest, it is often of interest to find the forces exerted by the fluid on a solid surface. The mechanics of a stationary fluid, too, is indeed governed by the incompressible Navier–Stokes equation set (5.110). The velocity field, being identically zero everywhere, implies that the velocity gradient tensor, too, is identically zero everywhere. Thus, the continuity equation $\underline{\nabla} \cdot \underline{V} = 0$ is trivially satisfied. On the other hand, in the governing equation of $\underline{V}$ (5.110), the unsteady term, the advection term, and the viscous term are zero, leading to the following simplified form of the equation.

$$0 = -\underline{\nabla} p + \rho \underline{g}. \tag{7.1}$$

This equation shows that every fluid particle is in a state of equilibrium, and the net external force arising due to pressure is balanced by the net gravitational force acting on it. In its current form, the simplified Eq. (7.1) is still a differential equation with p being the only unknown. However, this equation is now simple enough for us to determine its exact solution. Let us employ a frame-fixed Cartesian coordinate system $Ox_1(\hat{e}_1)x_2(\hat{e}_2)x_3(\hat{e}_3)$ as our working coordinate system, and without any loss of generality, let us orient the axes of the coordinate system such that $\underline{g} = -g\hat{e}_2$. Equation (7.1) is then expressed as:

$$\frac{1}{\rho}\frac{\partial p}{\partial x_1}\hat{e}_1 + \frac{1}{\rho}\frac{\partial p}{\partial x_2}\hat{e}_2 + \frac{1}{\rho}\frac{\partial p}{\partial x_3}\hat{e}_3 = -g\hat{e}_2. \tag{7.2}$$

This is a set of three scalar differential equations, which are cast as

$$\frac{1}{\rho}\frac{\partial p}{\partial x_1} = 0, \ \frac{1}{\rho}\frac{\partial p}{\partial x_2} = -g, \ \frac{1}{\rho}\frac{\partial p}{\partial x_3} = 0. \tag{7.3}$$

© The Author(s), under exclusive license to Springer Nature Switzerland AG 2026
S. S. Sinha, *Fundamentals of Fluid Mechanics*,
https://doi.org/10.1007/978-3-032-04210-1_7

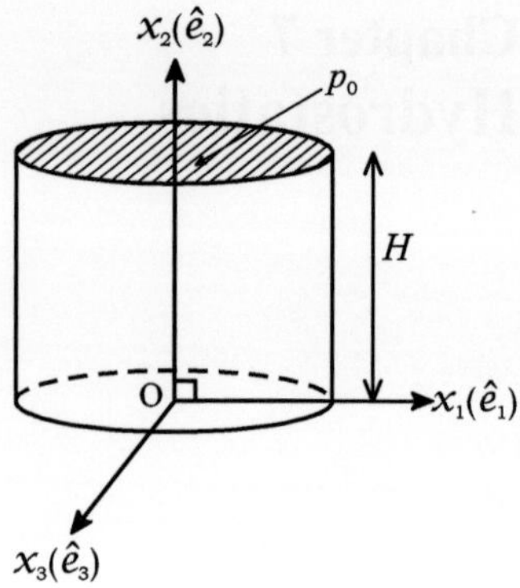

Fig. 7.1 A column of stationary fluid

In Fig. 7.1, we show a column of stationary liquid in a container that is open to the atmosphere and has pressure (p_o) at $x_2 = H$. The shaded area in the figure represents the free-surface of the fluid. The height of the container is H. Thus, the relevant boundary condition for the set of equations in (7.3) is

$$p(x_1, x_2, x_3 = H) = p_o \text{ for all } x_1 \text{ and } x_2. \tag{7.4}$$

Using the boundary condition (7.4), the exact general solution of (7.3) is

$$p(x_1, x_2, x_3) = p_o + \rho g(H - x_2). \tag{7.5}$$

Evidently, the solution (7.5) shows that (i) pressure within the fluid remains constant at a fixed height (x_2), and (ii) pressure increases linearly from the top ($x_2 = H$) of the container to its base ($x_2 = 0$). Such a pressure distribution is called the *hydrostatic* pressure distribution. If A and B are two arbitrary locations in the fluid domain, then (7.5) leads to the simple conclusion that the difference in pressure between the two locations is proportional to the difference in the heights of the two points.

$$p_B - p_A = \rho g \left[(x_2)_A - (x_2)_B\right], \tag{7.6}$$

where $(x_2)_A$ and $(x_2)_B$ are the heights of the two locations (A and B). Further, in the special case when there is no body force present ($g = 0$), (7.6) leads to the conclusion that $p_B = p_A$ (the pressure distribution is uniform across the fluid domain).

7.1 Net Hydrostatic Force on a Wetted Solid Surface

In Fig. 7.2, we show a solid surface submerged in a stationary liquid with its free surface at $x_2 = H$, and the pressure at that height is p_o. The acceleration due to the

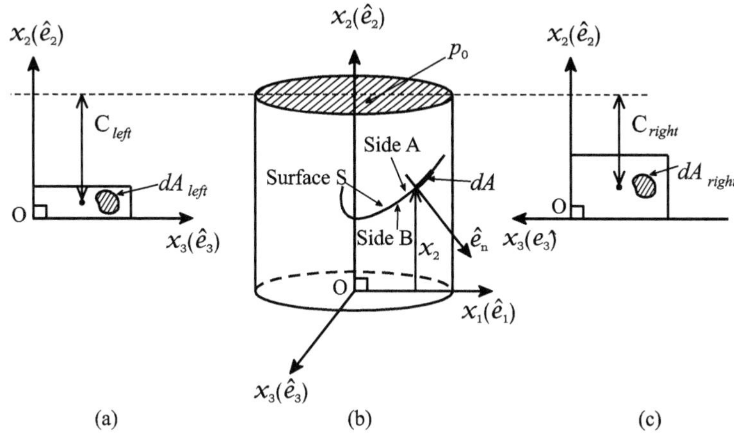

Fig. 7.2 A solid surface S submerged in a stationary fluid (sub-figure b). The sub-figures a and c show the projections A_{left} and A_{right} of surface S, respectively

gravity vector is oriented along $-\hat{e}_2$. The external force system acting on the solid surface is a surface-distributed force system arising from the local pressure stress tensor at the fluid–solid interface. We wish to find the resultant fluid force that acts specifically on side B of the surface (Fig. 7.2).

We focus our attention on a small area dA on side B (Fig. 7.2) with its centroid located at point Q having coordinates (x_1, x_2, x_3). The unit vector $\hat{e}_n$ represents the local outward normal on the surface. The force exerted by the fluid on the area dA of the solid surface is (expressed using the frame-fixed Cartesian coordinate system shown in Fig. 7.2)

$$\begin{aligned} d\underline{F} &= -pdA\hat{e}_n \\ &= -\left[pdA\hat{e}_n \cdot \hat{e}_1\right]\hat{e}_1 - \left[pdA\hat{e}_n \cdot \hat{e}_2\right]\hat{e}_2 - \left[pdA\hat{e}_n \cdot \hat{e}_3\right]\hat{e}_3. \end{aligned} \tag{7.7}$$

To find the net force $\underline{F}$ on the entire side B of the surface, we integrate (7.7) over the entire side B of the surface (S).

$$\begin{aligned} \underline{F} &= F_1\hat{e}_1 + F_2\hat{e}_2 + F_3\hat{e}_3 = \iint_S d\underline{F} \\ &= -\iint_S pdA\hat{e}_n \\ &= -\hat{e}_1\iint_S p\hat{e}_n \cdot \hat{e}_1 dA - \hat{e}_2\iint_S p\hat{e}_n \cdot \hat{e}_2 dA - \hat{e}_3\iint_S p\hat{e}_n \cdot \hat{e}_3 dA. \end{aligned} \tag{7.8}$$

Using the hydrostatic pressure distribution (7.5) and the available details about the variation of $\hat{e}_n$ over the solid surface, the integrals in (7.8) can be computed to find the components of $\underline{F}$ along the three Cartesian unit vectors. Alternatively, these integrals may also be computed in a simpler manner by not performing the integration on the actual wetted surface S, but on the appropriate projections of S on mutually perpendicular planes of the working Cartesian coordinate system. Depending on the sign of $\hat{e}_n \cdot \hat{e}_1$ the force along the $\hat{e}_1$ direction can be determined by performing two separate integrals over two different projected areas A_{left} and A_{right}.

$$F_1 = \iint_{A_{left}} p dA_{left} - \iint_{A_{right}} p dA_{right}, \tag{7.9}$$

where A_{left} and A_{right} are two separate projections on the $x_2(\hat{e}_2) - x_3(\hat{e}_3)$ plane of those parts of S where $\hat{e}_n \cdot \hat{e}_1 < 0$ and $\hat{e}_n \cdot \hat{e}_1 > 0$, respectively. The symbols dA_{left} and dA_{right} are infinitesimal local areas on the projected areas A_{left} and A_{right}, respectively (Fig. 7.2). Using (7.6) in (7.9) leads to

$$\begin{aligned} F_1 = p_o(A_{left} - A_{right}) + \rho g \iint_{A_{left}} (H - x_2) dA_{left} \\ - \rho g \iint_{A_{right}} (H - x_2) dA_{right}. \end{aligned} \tag{7.10}$$

If the symbols c_{left} and c_{right} denote the respective distances of the centroids of the areas A_{left} and A_{right}, respectively, measured from the free surface of the fluid along the direction of $\hat{e}_2$ (Fig. 7.2), then (7.10) can as well be expressed as:

$$F_1 = p_o(A_{left} - A_{right}) + \rho g c_{left} A_{left} - \rho g c_{right} A_{right}. \tag{7.11}$$

Similarly, F_3 can be computed as

$$F_3 = p_o(A_{rear} - A_{front}) + \rho g c_{rear} A_{rear} - \rho g c_{front} A_{front}, \tag{7.12}$$

where A_{rear} and A_{front} are separate projections on the $x_1(\hat{e}_1) - x_2(\hat{e}_2)$ plane of those parts of S where $\hat{e}_n \cdot \hat{e}_3 < 0$ and $\hat{e}_n \cdot \hat{e}_3 > 0$, respectively. The symbols dA_{rear} and dA_{front} are infinitesimal local areas on A_{rear} and A_{front}, respectively. the symbols c_{rear} and c_{front} denote the respective distances of the centroids of the areas A_{rear} and A_{front}, respectively, measured from the free surface of the fluid along the direction of $\hat{e}_2$. The expression for F_2 in (7.8) can alternatively be expressed as

$$F_2 = \iint_{A_{bottom}} p dA_{bottom} - \iint_{A_{top}} p dA_{top}, \tag{7.13}$$

where A_{bottom} and A_{top} are separate projections on the $x_3(\hat{e}_3) - x_1(\hat{e}_1)$ plane of those parts of S where $\hat{e}_n \cdot \hat{e}_2 < 0$ and $\hat{e}_n \cdot \hat{e}_2 > 0$, respectively. The symbols dA_{bottom} and dA_{top} are infinitesimal local areas on A_{bottom} and A_{top}, respectively. Using (7.6) in (7.13) leads to,

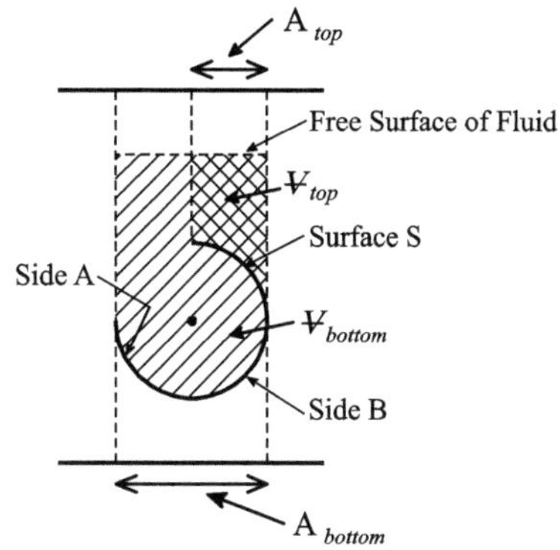

Fig. 7.3 Visualization of $\forall_{top}$ and $\forall_{bottom}$ with reference to side B of the solid surface S

$$F_2 = p_o(A_{bottom} - A_{top}) + \rho g \iint_{A_{bottom}} (H - x_2) dA_{bottom} - \rho g \iint_{A_{top}} (H - x_2) dA_{top}. \tag{7.14}$$

We recognize that the two integrals on the rhs of (7.14) represent some volumes.

$$\iint_{A_{bottom}} (H - x_2) dA_{bottom} = \forall_{bottom} \text{ and } \iint_{A_{top}} (H - x_2) dA_{bottom} = \forall_{top}, \tag{7.15}$$

where $\forall_{bottom}$ equals the volume of the three-dimensional space that exists between the free surface of the liquid and those parts of the surface area S where $\hat{e}_n \cdot \hat{e}_2 < 0$. Similarly, $\forall_{top}$ equals the volume of the three-dimensional space that exists between the free surface of the liquid and those parts of the surface area S where $\hat{e}_n \cdot \hat{e}_2 > 0$ (in Fig. 7.3 we show a typically shaped surface for illustration). Thus, the net force component F_2 (7.13) equals

$$F_2 = p_o(A_{bottom} - A_{top}) + \rho g \forall_{bottom} - \rho g \forall_{top}. \tag{7.16}$$

7.2 Resultant Hydrostatic Force System on a Submerged Solid Body

Figure 7.4 shows a solid body fully submerged in a stationary fluid. The free surface of the stationary fluid is at $x_2 = H$, and the pressure at that height is p_o. The acceleration due to the gravity vector is oriented along $-\hat{e}_2$. Again, the fluid exerts a distributed force system arising due to the pressure stress tensor existing on the solid–fluid

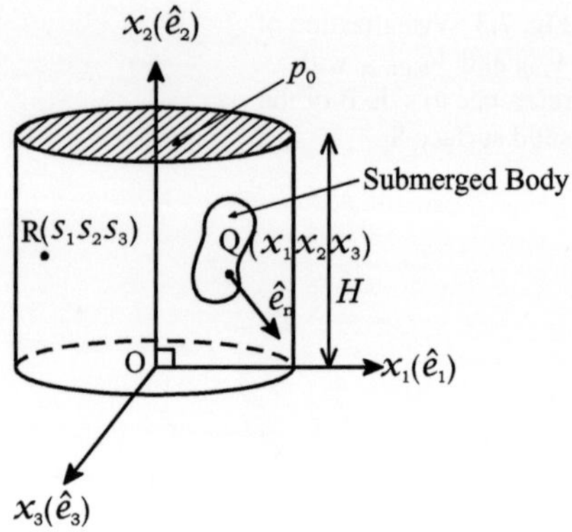

Fig. 7.4 A solid body submerged in a stationary fluid. Q is an arbitrary point on the surface of the solid object and has coordinates (x_1, x_2, x_3). $\hat{e}_n$ is the local outward unit normal vector

interface. We wish to find the resultant force system exerted by the fluid on the entire body.

The resultant force vector ($\underline{F}^{body}$) acting on the body is calculated by performing the following integral over the entire external closed surface (S) of the body,

$$\begin{aligned} \underline{F}^{body} &= \iint_S d\underline{F} = -\iint_S p dA \hat{e}_n = \\ &-\hat{e}_1 \iint_S p(\hat{e}_n \cdot \hat{e}_1) dA - \hat{e}_2 \iint_S p(\hat{e}_n \cdot \hat{e}_2) dA \\ &-\hat{e}_3 \iint_S p(\hat{e}_n \cdot \hat{e}_3) dA, \end{aligned} \tag{7.17}$$

where $\hat{e}_n$ represents the local outward normal unit vector and dA is an infinitesimal area on the solid surface. At all locations, $\hat{e}_n$ points away from the outer surface of the body.

For every arbitrarily chosen projected area $dA\hat{e}_n \cdot \hat{e}_1$ (an infinitesimal projection of the body surface area on a plane with its normal along $\hat{e}_1$) where the pressure is $p_o + \rho g(H - x_2)$, we can find a corresponding area with the same x_2 coordinate where the pressure force component along $\hat{e}_1$ would have identical magnitude but opposite sign. Further, these two force components would have a common line of action. While the algebraic sum of such a pair of force components would lead to no net force being experienced by the body in the $\hat{e}_1$ direction, the collinearity of such a pair implies that the net moment vector (about any point in space) due to these force components would be zero. Thus, the first integral on the rhs of (7.17) must vanish. Following a similar set of arguments, we conclude that the net force on the body along $\hat{e}_2$, the direction (the third integral in (7.17) must also vanish. Further, the net moment vector about any point arising because of the component of all the distributed pressure forces along the $\hat{e}_2$ direction must also be zero. The vanishing of the first and third integral in (7.8) leads to

$$\underline{F}^{body} = -\iint_S pdA(\hat{e}_n \cdot \hat{e}_2)\hat{e}_2. \tag{7.18}$$

Further, the second integral gets simplified to:

$$\underline{F}^{body} = \hat{e}_2 \iint_{A_{bottom}} pdA_{bottom} - \hat{e}_2 \iint_{A_{top}} pdA_{top}, \tag{7.19}$$

where A_{bottom} and A_{top} are separate projections on the $x_3(\hat{e}_3) - x_1(\hat{e}_1)$ plane of those parts of S where $\hat{e}_n \cdot \hat{e}_2 < 0$ and $\hat{e}_n \cdot \hat{e}_2 > 0$, respectively. The symbols dA_{bottom} and dA_{top} are infinitesimal local areas on A_{bottom} and A_{top}, respectively. Using the hydrostatic pressure distribution of (7.5), (7.19) is expressed as

$$F_2 = p_o(A_{bottom} - A_{top}) + \rho g \forall_{bottom} - \rho g \forall_{top}, \tag{7.20}$$

where $\forall_{bottom}$ equals the volume of the three-dimensional space that exists between the free surface of the fluid and those parts of the surface area S where $\hat{e}_n \cdot \hat{e}_2 < 0$. Similarly, $\forall_{top}$ equals the volume of the three-dimensional space that exists between the free surface of the fluid and those parts of the surface area S where $\hat{e}_n \cdot \hat{e}_2 > 0$. Since a submerged body has a closed wetted surface,

1. $A_{bottom} = A_{top}$,
2. $\forall_{body} = \forall_{bottom} - \forall_{top}$, where $\forall_{body}$ is the volume of the solid body.

Thus, (7.20) is expressed as

$$\underline{F}^{body} = \rho g \forall_{body}\hat{e}_2. \tag{7.21}$$

This net vertical force (7.21) acting on a submerged body is commonly referred to as the *buoyancy force* . Since $\rho g \forall_{body}$ equals the weight of the liquid displaced by the solid object, the magnitude of the buoyant force is often reported to be the *weight of the fluid displaced by the body*. As clearly demonstrated by our derivation in this section, the origin of this force is the hydrostatic pressure stress distribution existing on the solid-body interface (7.5). This pressure distribution, in turn, is the exact solution of the simplified form of the incompressible Navier–Stokes equation set.

If we wish to replace the distributed force system on a submerged body by a single force without any associated pure moment, then (i) that single force vector must equal $\underline{F}^{body}$ (7.21), and (ii) the line of action of that single force vector must pass through a special location R such that the net moment due to the distributed force system acting on the solid surface must be zero about that location (R). If location R has the coordinates (s_1, s_2, s_3) with reference to our working Cartesian coordinate system (Fig. 7.4), then these coordinates must satisfy the following condition:

$$\iint_S \{\underline{r}_{QR} \times [-pdA\hat{e}_n]\} = 0\,, \text{ or}$$

$$\iint_S \{(\underline{r}_{QO} - \underline{r}_{RO}) \times [-pdA\hat{e}_n]\} = 0\,, \text{or}$$
$$\iint_S \{[(x_1 - s_1)\hat{e}_1 + (x_2 - s_2)\hat{e}_2 + (x_3 - s_3)\hat{e}_3] \times [-pdA\hat{e}_n]\} = 0\,, \text{or}$$
$$\begin{aligned}&\iint_S \{[(x_1 - s_1)\hat{e}_1 + (x_2 - s_2)\hat{e}_2 + (x_3 - s_3)\hat{e}_3] \times [-pdA(\hat{e}_n \cdot \hat{e}_1)\hat{e}_1]\}\\ &+ \iint_S \{[(x_1 - s_1)\hat{e}_1 + (x_2 - s_2)\hat{e}_2 + (x_3 - s_3)\hat{e}_3] \times [-pdA(\hat{e}_n \cdot \hat{e}_2)\hat{e}_2]\}\\ &+ \iint_S \{[(x_1 - s_1)\hat{e}_1 + (x_2 - s_2)\hat{e}_2 + (x_3 - s_3)\hat{e}_3] \times [-pdA(\hat{e}_n \cdot \hat{e}_3)\hat{e}_3]\}\\ &= 0,\end{aligned} \tag{7.22}$$

where Q with coordinates (x_1, x_2, x_3) is an arbitrary point on the surface of the solid body, and $\hat{e}_n$ is the local outward unit normal vector at Q. Earlier in this section, we have highlighted that the collinearity of the pair-wise infinitesimal force components along the $\hat{e}_1$ and $\hat{e}_2$ directions ensure that their net moments about any point in space must vanish. Thus, the first and third integral on the left-hand side (lhs) of (7.22) must vanish, resulting in the following condition on (s_1, s_2, s_3).

$$\begin{aligned}&\iint_S \{[(x_1 - s_1)\hat{e}_1 + (x_2 - s_2)\hat{e}_2 + (x_3 - s_3)\hat{e}_3] \times [-pdA(\hat{e}_n \cdot \hat{e}_2)\hat{e}_2]\} = 0,\\ &\Rightarrow -\hat{e}_3 \iint_S (x_1 - s_1)pdA(\hat{e}_n \cdot \hat{e}_2) + \hat{e}_1 \iint_S (x_3 - s_3)pdA(\hat{e}_n \cdot \hat{e}_2) = 0,\end{aligned} \tag{7.23}$$

which leads to two scalar equations

$$\begin{aligned}&\iint_S (x_1 - s_1)pdA(\hat{e}_n \cdot \hat{e}_2) = 0,\\ &\iint_S (x_3 - s_3)pdA(\hat{e}_n \cdot \hat{e}_2) = 0.\end{aligned} \tag{7.24}$$

Re-arranging the equations of (7.24) leads us to

$$s_1 = \frac{\iint_S x_1 pdA(\hat{e}_n \cdot \hat{e}_2)}{\iint_S pdA(\hat{e}_n \cdot \hat{e}_2)} = \frac{\iint_{A_{top}} x_1 pdA_{top} - \iint_{A_{bottom}} x_1 pdA_{bottom}}{\iint_{A_{top}} pdA_{top} - \iint_{A_{bottom}} pdA_{bottom}},$$

$$s_3 = \frac{\iint_S x_3 pdA(\hat{e}_n \cdot \hat{e}_2)}{\iint_S pdA(\hat{e}_n \cdot \hat{e}_2)} = \frac{\iint_{A_{top}} x_3 pdA_{top} - \iint_{A_{bottom}} x_3 pdA_{bottom}}{\iint_{A_{top}} pdA_{top} - \iint_{A_{bottom}} pdA_{bottom}}, \tag{7.25}$$

where A_{bottom} and A_{top} are the same two separate projections on the $x_3(\hat{e}_3) - x_1(\hat{e}_1)$ plane of those parts of S where $\hat{e}_n \cdot \hat{e}_2 < 0$ and $\hat{e}_n \cdot \hat{e}_2 > 0$, respectively. The symbols dA_{bottom} and dA_{top} are infinitesimal local areas on A_{bottom} and A_{top}, respectively. For the submerged solid body,

1. $A_{bottom} = A_{top}$.

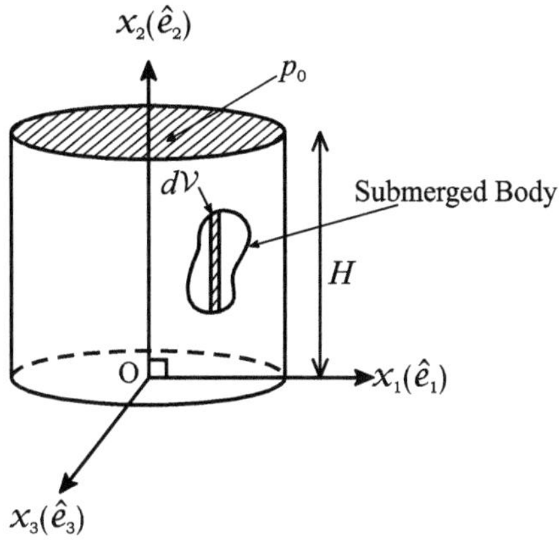

Fig. 7.5 An infinitesimal slender solid volume $d\mathcal{V}$ having a finite length along the direction of $\hat{e}_2$ with a uniform cross section area dA_2. This slender volume is oriented along the direction of $\hat{e}_2$

2. For every selected infinitesimal area dA_{bottom} located at $(x_1,(x_2)_{bottom},x_3)$, there must exist another corresponding infinitesimal area dA_{top} located at $(x_1,(x_2)_{top},x_3)$ such that $dA_{top} = dA_{bottom}$.

Thus, the two separate integrals can be performed in a combined manner as

$$\iint_{A_{top}} x_1 p dA_{top} - \iint_{A_{bottom}} x_1 p dA_{bottom} = \iint_{A_{bottom}} x_1 \left[p_{top} - p_{bottom}\right] dA_{bottom}. \tag{7.26}$$

At the centroid of these corresponding areas dA_{top} and dA_{bottom}, the difference in the pressure values is expressed using the hydrostatic pressure distribution (7.6)

$$p_{top} - p_{bottom} = \rho g \left[(x_2)_{bottom} - (x_2)_{top}\right]. \tag{7.27}$$

Using (7.27) in (7.26) leads to

$$\iint_{A_{top}} x_1 p dA_{top} - \iint_{A_{bottom}} x_1 p dA_{bottom} = \rho g \iint_{A_{bottom}} x_1 \left[(x_2)_{bottom} - (x_2)_{top}\right] dA_{bottom}. \tag{7.28}$$

The product $\left[(x_2)_{bottom} - (x_2)_{top}\right] dA_{bottom}$ represents the volume $(d\mathcal{V})$ of a slender element of the solid body as shown in Fig. 7.5. This volume is oriented along the direction of $\hat{e}_2$. Thus, (7.28) can be expressed as a volume integral,

$$\iint_{A_{top}} x_1 p dA_{top} - \iint_{A_{bottom}} x_1 p dA_{bottom}$$

$$= \rho g \iint_{A_{bottom}} x_1 \left[(x_2)_{bottom} - (x_2)_{top} \right] dA_{bottom},$$
$$= \rho g \iiint_{\mathcal{V}} x_1 d\mathcal{V}, \tag{7.29}$$

where the subscript the symbol $\iiint_{\mathcal{V}}$ represents integration over the entire volume of the solid body. Following a similar set of steps, we convert the other area integrals of (7.25) into volume integrals,

$$\iint_{A_{top}} p dA_{top} - \iint_{A_{bottom}} p dA_{bottom} = \rho g \iiint_{\mathcal{V}} d\mathcal{V} = \rho g \mathcal{V}_{body},$$
$$\iint_{A_{top}} x_3 p dA_{top} - \iint_{A_{bottom}} x_3 p dA_{bottom} = \rho g \iiint_{\mathcal{V}} x_3 d\mathcal{V}, \tag{7.30}$$

where $\mathcal{V}_{body}$ represents the volume of the submerged solid body. Using (7.30) in (7.25) leads to,

$$s_1 = \frac{\iiint_{\mathcal{V}} x_1 d\mathcal{V}}{\mathcal{V}_{body}},$$
$$s_3 = \frac{\iiint_{\mathcal{V}} x_3 d\mathcal{V}}{\mathcal{V}_{body}}. \tag{7.31}$$

By definition, the right-hand sides of the two equations in (7.31) are the first and third coordinates of the location of the centroid of the volume of the submerged solid body. Clearly, (7.31) imposes constraints on s_1 and s_3 but not on s_2. Thus, any point on the line parallel to $\hat{e}_2$ and passing through the centroid of the solid volume is a legitimate point where the net force vector $\underline{F}^{body}$ (7.21) can be placed (without any associated pure moment) to arrive at a force system which is equivalent to the actual distributed pressure force system acting on the submerged solid body.

Chapter 8
Some Simple Representative Flows of an Ideal Fluid

In this chapter, we study some simple flow fields of ideal fluids. The stress tensor in such flow fields is entirely due to pressure. There is no viscous stress tensor present. Further, the velocity field is not subject to the no-slip condition on solid interfaces. However, the no-penetration condition still applies.

Ideal fluids are not found in nature. However, there are reasons why a good understanding of the behavior of ideal fluid flows is still useful. For ideal fluids, we can simplify the governing Navier–Stokes equation to the extent that for some simple flow fields, we can obtain exact solutions for the pressure and the velocity fields. Such solutions can then be used to develop insights into some aspects of flow physics. Specifically, if available, such an exact solution of a flow past a solid body can help us understand how the no-penetration condition imposed by the solid body influences the pressure and velocity fields in its vicinity. Further, it turns out (more on this in Chap. 9) that, even with real fluids under certain conditions, the behavior of the flow in some regions of the flow field turns out to be approximately the same as the solution of a flow field with an ideal fluid.

8.1 Governing Equations of an Ideal Fluid

Setting $\mu = 0$ in the incompressible Navier–Stokes equation set (5.110) leads to its following simplified form which is applicable to the flow of ideal fluids

$$\begin{aligned} \underline{\nabla} \cdot \underline{V} &= 0, \\ \rho \frac{\partial \underline{V}}{\partial t} + \rho \left(\underline{V} \cdot \underline{\nabla} \right) \underline{V} &= -\underline{\nabla} p + \rho \underline{g}. \end{aligned} \tag{8.1}$$

© The Author(s), under exclusive license to Springer Nature Switzerland AG 2026

S. S. Sinha, *Fundamentals of Fluid Mechanics*,
https://doi.org/10.1007/978-3-032-04210-1_8

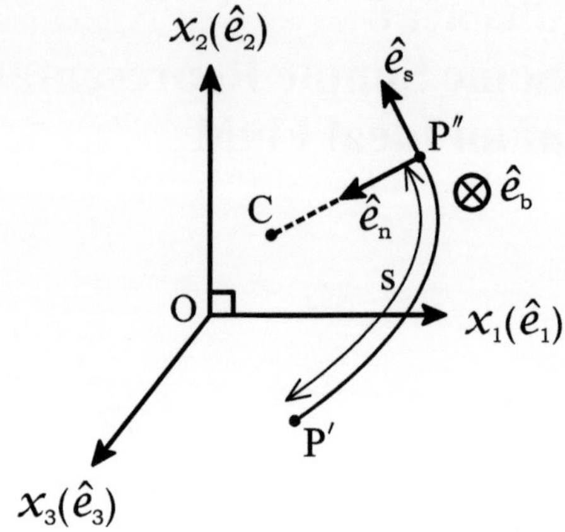

Fig. 8.1 A streamline and the description of the path coordinate system

The difference between (8.1) and (5.110) is the absence of the viscous force term on the rhs of the governing equation of $\underline{V}$. Equation set (8.1) is commonly referred to as the *incompressible Euler equation* set.

8.2 Bernoulli's Equation

For a steady state velocity field, the partial differential equation governing the velocity vector in the Euler equation set (8.1) can be significantly simplified. Actually, it is possible to appropriately integrate this equation and transform it into an algebraic equation relating the magnitude of the velocity vector to the local pressure. For this simplification, however, we choose a path coordinate system as our working coordinate system.

In Fig. 8.1, we show a representative streamline curve ($P^{'}P^{''}$) in a flow field having a steady velocity field. The fluid particle which is currently present at the location P'' is denoted by P. The location of particle P at a reference time in the past was $P^{'}$. Since the velocity field is steady, the streamline curve $P^{'}P^{''}$ is also the path-line of the fluid particle P. The symbol s in the figure represents the current distance of the particle P measured from $P^{'}$ along the streamline. Further, two unit vectors are also shown in Fig. 8.1. The unit vector $\hat{e}_s$ points along the instantaneous velocity vector of the fluid particle P. Thus, $\hat{e}_s$ points along the local tangent of the path-line at location $P^{'}$ (it is referred to as the *local tangent unit vector*). The unit vector $\hat{e}_n$ is chosen to be such that

$$\hat{e}_n = \frac{\partial \hat{e}_s/\partial s}{|\partial \hat{e}_s/\partial s|}, \tag{8.2}$$

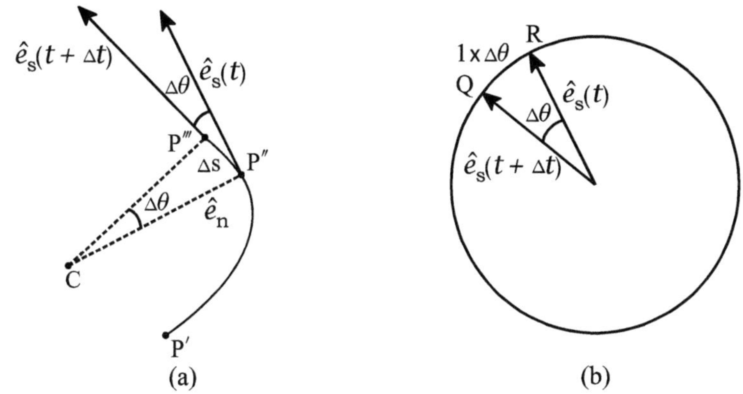

Fig. 8.2 Geometrical details about the unit vectors of the path coordinate system at two time instants

where $|.|$ means the magnitude of the vector in context. It can be shown that $\hat{e}_n$ is perpendicular to $\hat{e}_s$:

$$\hat{e}_s \cdot \hat{e}_s = 1 \implies \frac{\partial \left(\hat{e}_s \cdot \hat{e}_s\right)}{\partial s} = 0 \implies 2\hat{e}_s \cdot \frac{\partial \hat{e}_s}{\partial e_s} = 0,$$
$$\implies 2\hat{e}_s \cdot \hat{e}_n |\partial \hat{e}_s / \partial s| = 0 \implies \hat{e}_s \cdot \hat{e}_n = 0\,,\ \text{if}\ |\partial \hat{e}_s/\partial s| \neq 0. \tag{8.3}$$

The unit vector $\hat{e}_n$ is called the *principal normal unit vector*. To ascertain the magnitude, $|\partial \hat{e}_s / \partial s|$ we refer to Fig. 8.2a. In this figure, we show tangent unit vectors at two time instants, which differ infinitesimally. The unit vector $\hat{e}_s(t)$ is the unit vector at time t, the unit vector $\hat{e}_s(t + \Delta t)$ is the tangent unit vector at time $t + \Delta t$. At time t the fluid particle P is at location $P^{''}$ and at time $t + \Delta t$ the fluid particle is at location $P^{'''}$. The distance between these locations along the streamline is Δs. If R is the radius of curvature of the streamline curve at location P', and C represents the corresponding center of curvature, then the geometry of the triangle $CP^{''}P^{'''}$ ensures that

$$\Delta s \approx R \Delta\theta, \tag{8.4}$$

where $\Delta\theta$ is the infinitesimal angle by which the orientation of the two unit vectors differs.

In Fig. 8.2b, the same two unit vectors have been shown with their tails coinciding with each other without altering their directions. Since both these vectors have equal magnitude (unity), their arrowheads must lie on a unit circle, with the center being the point where their tails lie (Fig. 8.2b). Further, the angle between the two vectors must be $\Delta\theta$. Referring to this unit circle, the length of the displacement vector $|\underline{r}_{QR}|$ is approximated as

$$|\underline{r}_{QR}| \approx |\hat{e}_s(t + \Delta t) - \hat{e}_s(t)| \approx 1 \times \Delta\theta. \tag{8.5}$$

Using (8.4) and (8.5) we are now in a position to find $|\partial \hat{e}_s/\partial s|$:

$$|\frac{\partial \hat{e}_s}{\partial s}| = \lim_{\Delta s \to 0} \frac{|\hat{e}_s(t+\Delta t) - \hat{e}_s(t)|}{\Delta s} = \lim_{\Delta s \to 0} \frac{1 \times \Delta\theta}{R\Delta\theta} = \frac{1}{R}. \tag{8.6}$$

Thus, using (8.2) and (8.6) we conclude

$$\hat{e}_n = R\frac{\partial \hat{e}_s}{\partial s}. \tag{8.7}$$

Further, we define another unit vector $\hat{e}_b$ (called the bi-normal vector) which is defined as $\hat{e}_b = \hat{e}_s \times \hat{e}_n$. The path coordinate system employs the scalar quantity s to ascertain the instantaneous location of the particle P, whereas the set of three mutually perpendicular unit vectors $\hat{e}_s$, $\hat{e}_n$, and $\hat{e}_b$ is used to express various tensors in this coordinate system. Owing to the ways in which the three unit vectors are defined, their directions depend on the instantaneous location of the particle P. Thus, these unit vectors change as the particle P moves in the flow field.

At the current instant, the velocity of particle P is expressed in the path coordinate system as,

$$\underline{V} = V\hat{e}_s, \tag{8.8}$$

where V represents the speed of the particle. Further, in the path coordinate system, the nabla ($\underline{\nabla}$) operator is expressed as:

$$\underline{\nabla} = \hat{e}_s\frac{\partial}{\partial s} + \hat{e}_n\frac{\partial}{\partial n} + \hat{e}_b\frac{\partial}{\partial b}, \tag{8.9}$$

where the partial derivatives appearing on the rhs of (8.9) are to be computed by allowing infinitesimal independent changes Δs, Δn and Δb to the current spatial location of the particle P along the directions of the instantaneous unit vectors $\hat{e}_s$, $\hat{e}_n$ and $\hat{e}_b$, respectively. Using (8.8) and (8.9) the governing equation of $\underline{V}$ in (8.1) is expressed as

$$\left[V\hat{e}_s \cdot \left(\hat{e}_s\frac{\partial}{\partial s} + \hat{e}_n\frac{\partial}{\partial n} + \hat{e}_b\frac{\partial}{\partial b}\right)\right] V\hat{e}_s = -\frac{1}{\rho}\left(\hat{e}_s\frac{\partial}{\partial s} + \hat{e}_n\frac{\partial}{\partial n} + \hat{e}_b\frac{\partial}{\partial b}\right) p + \underline{g}. \tag{8.10}$$

Equation (8.10), when expanded, leads to the following equation:

$$\left[V\hat{e}_s \cdot \hat{e}_s\frac{\partial}{\partial s} + V\hat{e}_s \cdot \hat{e}_n\frac{\partial}{\partial n} + V\hat{e}_s \cdot \hat{e}_b\frac{\partial}{\partial b}\right](V\hat{e}_s) = -\frac{1}{\rho}\left(\hat{e}_s\frac{\partial p}{\partial s} + \hat{e}_n\frac{\partial p}{\partial n} + \hat{e}_b\frac{\partial p}{\partial b}\right) + \underline{g}. \tag{8.11}$$

Since $\hat{e}_s$, $\hat{e}_n$, and $\hat{e}_b$ are mutually perpendicular, (8.10) simplifies to:

$$\left(V\frac{\partial}{\partial s}\right)(V\hat{e}_s) = V\frac{\partial V}{\partial s}\hat{e}_s + V^2\frac{\partial \hat{e}_s}{\partial s}$$
$$= -\frac{1}{\rho}\left(\hat{e}_s\frac{\partial p}{\partial s} + \hat{e}_n\frac{\partial p}{\partial n} + \hat{e}_b\frac{\partial p}{\partial b}\right) + \underline{g}. \tag{8.12}$$

Further,

$$V\frac{\partial V}{\partial s}\hat{e}_s = \frac{\partial}{\partial s}\left(\frac{V^2}{2}\right)\hat{e}_s. \tag{8.13}$$

Using (8.13) and (8.7) in (8.12) we arrive at the following form of the governing equation of the velocity vector:

$$\frac{\partial}{\partial s}\left(\frac{V^2}{2}\right)\hat{e}_s + \frac{V^2}{R}\hat{e}_n =$$
$$-\frac{1}{\rho}\frac{\partial p}{\partial s}\hat{e}_s - \frac{1}{\rho}\frac{\partial p}{\partial n}\hat{e}_n - \frac{1}{\rho}\frac{\partial p}{\partial b}\hat{e}_b + g_s\hat{e}_s + g_n\hat{e}_n + g_b\hat{e}_b. \tag{8.14}$$

In (8.14), we have expressed $\underline{g}$ also in the path coordinate system. The symbols g_s, g_n, and g_b are the scalar components of the vector $\underline{g}$ along the three unit vectors of the path coordinate system. The vector equation (8.14) can be equivalently expressed as a set of three independent scalar equations (along each of three unit vectors)

$$\frac{\partial}{\partial s}\left(\frac{V^2}{2}\right) = -\frac{1}{\rho}\frac{\partial p}{\partial s} + g_s,$$
$$\frac{V^2}{R} = -\frac{1}{\rho}\frac{\partial p}{\partial n} + g_n,$$
$$0 = -\frac{1}{\rho}\frac{\partial p}{\partial b} + g_b. \tag{8.15}$$

We focus our attention on the first equation of the set (8.15) and integrate it with respect to the path coordinate s from location $P^{'}$ to $P^{''}$ (Fig. 8.1).

$$\int_{P'}^{P''}\frac{\partial}{\partial s}\left(\frac{V^2}{2}\right)ds + \int_{P'}^{P''}\frac{1}{\rho}\frac{\partial p}{\partial s}ds = \int_{P'}^{P''} g_s ds. \tag{8.16}$$

The pressure and the velocity terms in (8.16) can be readily integrated as

$$\left(\frac{V^2}{2}+\frac{p}{\rho}\right)_{P''} - \left(\frac{V^2}{2}+\frac{p}{\rho}\right)_{P'} = \int_{P'}^{P''} g_s ds. \tag{8.17}$$

At this point, we assume that the vector $\underline{g}$ is oriented opposite to the unit vector $\hat{e}_2$, which is one of the unit vectors of the frame-fixed Cartesian coordinate system as shown in Fig. 8.1. Thus, the body force term in (8.17) can be alternatively expressed in the background Cartesian coordinate system as,

$$\begin{aligned} \int_{P'}^{P''} g_s ds &= \int_{P'}^{P''} \underline{g} \cdot d\underline{s}, \\ &= -\int_{P'}^{P''} g\hat{e}_2 \cdot d\underline{s}, \\ &= -\int_{P'}^{P''} g\hat{e}_2 \cdot \left(dx_1\hat{e}_1 + dx_2\hat{e}_2 + dx_3\hat{e}_3\right), \end{aligned} \tag{8.18}$$

where d$\underline{s}$ is the infinitesimal displacement vector along $\hat{e}_s(t)$. This infinitesimal vector can alternatively be expressed using the background frame-fixed Cartesian coordinate system using the scalar components dx_1, dx_2, and dx_3.

$$d\underline{s} = dx_1\hat{e}_1 + dx_2\hat{e}_2 + dx_3\hat{e}_3. \tag{8.19}$$

Thus, the rhs of (8.18) can be expressed as

$$\int_{P'}^{P''} g_s ds = -\int_{(x_2)_{P'}}^{(x_2)_{P''}} g dx_2 = -g\left[(x_2)_{P''} - (x_2)_{P'}\right], \tag{8.20}$$

where $(x_2)_{P''}$ and $(x_2)_{P'}$ are the x_2 coordinates of the locations P'' and P' in the background Cartesian coordinate system. Using (8.20) in (8.17) and subsequently re-arranging, we arrive at the fully integrated form of the equation

$$\left(\frac{V^2}{2g} + \frac{p}{\rho g} + x_2\right)_{P'} = \left(\frac{V^2}{2g} + \frac{p}{\rho g} + x_2\right)_{P''}. \tag{8.21}$$

Equation (8.21) is an algebraic equation relating pressure and the magnitude of velocity at two different locations P' and P'' which are lying on the same streamline. This equation is called *Bernoulli's equation.* In summary, this algebraic equation has been derived under the following set of conditions: (i) the density of the fluid is constant, (ii) the fluid is ideal, (iii) the velocity field is steady, (iv) the body force is constant and acts along the direction of $-\hat{e}_2$ and (v) the locations P' and P'' lie on the same streamline.

The quantities $V^2/2g$, $p/\rho g$, and x_2 appearing in Bernoulli's equation are called the local *velocity head*, the local *pressure head*, and the local *elevation head*, respectively. The sum of these three quantities is called the *total head* of the flow at a given location. Bernoulli's equation states that the total head of the flow along every streamline in a steady flow of an ideal, constant-density fluid remains constant. Bernoulli's equation may also be interpreted in terms of work and energy: *the gain in the kinetic energy of a fluid particle while it moves along a streamline in a steady flow field of an ideal, constant-density fluid equals the sum of the work done by the pressure force and the work done by the constant body force on that fluid particle*.

Even though Bernoulli's equation is an algebraic equation, by itself, it is not the complete solution of the Euler equation (8.1). The flow of an ideal fluid of constant density has four unknowns: the three components of the velocity vector and pressure. In the process of integrating the governing equation of $\underline{V}$, we have converted this vector equation into a single scalar equation (8.21), which merely relates the magnitude of the local velocity vector to the local pressure value. Even for those flow fields in which Bernoulli's equation is applicable, the solution of the three components of the velocity vector still remains to be determined.

8.3 The Vorticity Equation

In Chap. 4, we defined the vorticity vector ($\underline{\omega}$) to be the curl of the velocity field.

$$\underline{\omega} = \underline{\nabla} \times \underline{V}. \tag{8.22}$$

The vorticity vector represents twice the averaged rate of rotation of a small local fluid element (4.38) and thus has its own physical significance (4.38). At this point, we wish to derive the governing equation of vorticity more as a mathematical requirement in order to simplify the Euler equation set further. With such an equation, we wish to examine under what conditions the flow field of an ideal fluid is vorticity-free. If such conditions exist, we will show that the velocity vector can be expressed as the gradient of a scalar field. Such a representation of the velocity vector, in turn, will help us further simplify the steady Euler equations and take us closer to arriving at some exact solution of the velocity and pressure fields.

The process of deriving the governing equation of the vorticity vector begins by taking the curl of the governing equation of $\underline{V}$ in the Euler equation set (8.1)

$$\underline{\nabla} \times \left[\rho \frac{\partial \underline{V}}{\partial t} + \rho \left(\underline{V} \cdot \underline{\nabla}\right) \underline{V}\right] = \underline{\nabla} \times \left[-\underline{\nabla} p + \rho \underline{g}\right], \text{ or}$$

$$\underline{\nabla} \times \left[\rho \frac{\partial \underline{V}}{\partial t}\right] + \underline{\nabla} \times \left[\rho \left(\underline{V} \cdot \underline{\nabla}\right) \underline{V}\right] = \underline{\nabla} \times \left[-\underline{\nabla} p\right] + \left[\underline{\nabla} \times \left(\rho \underline{g}\right)\right]. \tag{8.23}$$

The curl of the constant gravitational acceleration must be zero

$$\underline{\nabla} \times \left(\rho \underline{g}\right) = 0. \tag{8.24}$$

Further, we have already seen that the curl of the gradient of even a spatially varying scalar field is zero (($\underline{\nabla} \times \left[-\underline{\nabla} p\right] = 0$), Eq. 1.62). Thus, the entire rhs of (8.23) is zero, and accordingly (8.23) simplifies to

$$\rho \frac{\partial \left[\underline{\nabla} \times \underline{V}\right]}{\partial t} + \rho \underline{\nabla} \times \left[\left(\underline{V} \cdot \underline{\nabla}\right) \underline{V}\right] = 0, \text{ or}$$

$$\rho \frac{\partial \underline{\omega}}{\partial t} + \rho \underline{\nabla} \times \left[\left(\underline{V} \cdot \underline{\nabla}\right) \underline{V}\right] = 0, \tag{8.25}$$

where ρ is a constant. Using the tensor identity (1.12), the second term on the lhs of (8.25) can be split into two terms

$$\begin{aligned} \underline{\nabla} \times \left[(\underline{V} \cdot \underline{\nabla})\underline{V}\right] &= \underline{\nabla} \times \left[\frac{\underline{\nabla}\left(\underline{V} \cdot \underline{V}\right)}{2} - \underline{V} \times \left(\underline{\nabla} \times \underline{V}\right)\right], \\ &= \underline{\nabla} \times \left[\frac{\underline{\nabla}\left(\underline{V} \cdot \underline{V}\right)}{2}\right] - \underline{\nabla} \times \left[\underline{V} \times \left(\underline{\nabla} \times \underline{V}\right)\right], \\ &= \underline{\nabla} \times \left[\frac{\underline{\nabla}\left(\underline{V} \cdot \underline{V}\right)}{2}\right] + \underline{\nabla} \times \left[\underline{\omega} \times \underline{V}\right]. \end{aligned} \tag{8.26}$$

The first term on the rhs is curl of gradient of a scalar $(\underline{V} \cdot \underline{V}/2)$, which is always zero (1.62). The second term on the rhs can be expanded using (1.68). Thus, (8.26) is modified to

$$\begin{aligned} \underline{\nabla} \times \left[(\underline{V} \cdot \underline{\nabla})\underline{V}\right] &= \underline{\nabla} \times \left(\underline{\omega} \times \underline{V}\right), \\ = \left[\left(\underline{\nabla} \cdot \underline{V}\right) \underline{\omega} - \left(\underline{\omega} \cdot \underline{\nabla}\right) \underline{V}\right] &+ \left[\left(\underline{V} \cdot \underline{\nabla}\right) \underline{\omega} - \left(\underline{\nabla} \cdot \underline{\omega}\right) \underline{V}\right]. \end{aligned} \tag{8.27}$$

For a fluid of constant density $\underline{\nabla} \cdot \underline{V} = 0$. Further, the identity (1.66) shows that $\underline{\nabla} \cdot \underline{\omega} = 0$. Thus, (8.27) simplifies to

$$\begin{aligned} \underline{\nabla} \times \left[(\underline{V} \cdot \underline{\nabla})\underline{V}\right] &= \underline{\nabla} \times \left(\underline{\omega} \times \underline{V}\right), \\ &= -\left(\underline{\omega} \cdot \underline{\nabla}\right) \underline{V} + \left(\underline{V} \cdot \underline{\nabla}\right) \underline{\omega}. \end{aligned} \tag{8.28}$$

Using (8.28) in (8.25) we arrive at the governing equation of $\underline{\omega}$

$$\begin{aligned} \frac{\partial \underline{\omega}}{\partial t} + \left(\underline{V} \cdot \underline{\nabla}\right) \underline{\omega} &= \frac{1}{\rho} \left(\underline{\omega} \cdot \underline{\nabla}\right) \underline{V}, \text{ or} \\ \frac{D\underline{\omega}}{Dt} &= \frac{1}{\rho} \left(\underline{\omega} \cdot \underline{\nabla}\right) \underline{V}. \end{aligned} \tag{8.29}$$

Equation (8.29) describes the change in the vorticity vector following the motion of a local fluid particle. It states that the cause of the change in the vorticity vector following a local fluid particle is the process represented mathematically by $\frac{1}{\rho} \left(\underline{\omega} \cdot \underline{\nabla}\right) \underline{V}$. This process is commonly referred to as the *vortex-stretching* process. This process has an explicit dependence on the local instantaneous vorticity vector. It is clear from (8.29) that if the $\underline{\omega}$ associated with an identified fluid particle was zero at a refer-

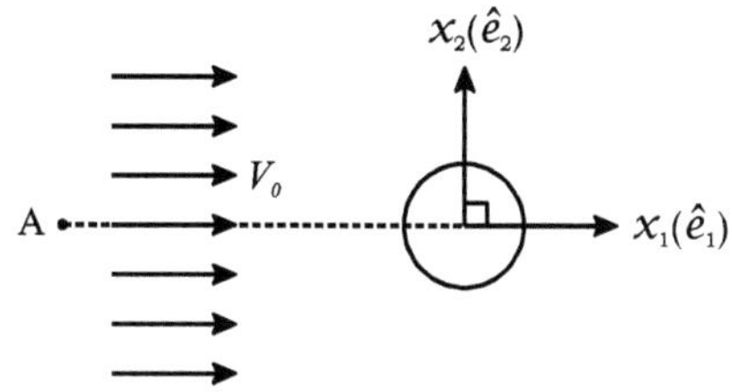

Fig. 8.3 A representative flow field past a cylinder (fixed to the ground frame) with a uniform velocity field in the far-upstream region

ence time in the past, it is guaranteed that the vortex-stretching process will never be activated and that fluid particle will continue to have zero $\underline{\omega}$ at all time instants.

8.4 Potential Flows

We now focus on a flow field shown in Fig. 8.3. The velocity field in the far-upstream region ($x_1 \to -\infty$ and at all x_2, x_3) is known to be uniform. This far-upstream condition is represented using a frame-fixed Cartesian coordinate system as

$$\underline{V} = V_o\hat{e}_1 \text{ as } x_1 \to -\infty \text{ at all } x_2, \text{ at all } x_3 \text{ and at all } t. \tag{8.30}$$

where V_o is a known scalar. Clearly, in the far-upstream region, $\underline{\omega} = \underline{\nabla} \times \underline{V}$ is zero. Thus, in the far-upstream region, all fluid particles have zero vorticity associated with them. If we identify any particle in the far-upstream region and then follow its motion, (8.29) guarantees that the associated vorticity must be zero irrespective of the location to which the particle moves at any later time. Since for a flow field as shown in the Fig. 8.3, every constituent fluid particle in the entire flow domain must have moved from its initial location in the far-upstream region, we conclude that every fluid particle present in such a flow field will remain free of vorticity at all time instants. For such a flow field, at all locations,

$$\underline{\omega} = 0. \tag{8.31}$$

Since the curl of the gradient of a scalar field is always zero (1.62), (8.31) is unconditionally satisfied by the following form of the velocity vector

$$\underline{V}(\underline{X}, t) = \underline{\nabla}\left[\phi\left(\underline{X}, t\right)\right], \tag{8.32}$$

where ϕ generally is a spatially- and time-varying Eulerian scalar field. The symbol $\underline{X}$ is the independent spatial variable of the Eulerian description of the flow field.

Even though at this stage we may not know the actual mathematical form of ϕ, nonetheless, it is a significant advancement in our quest to obtain some exact solutions of the Euler equation set (8.1) that all three components of the velocity field can be determined from a scalar field itself. The scalar field ϕ is called the *potential function* of the velocity field. A flow in which velocity can be expressed as the gradient of a scalar field is called a *potential flow*.

With the velocity being represented as in (8.32), both the continuity and the governing equation of $\underline{V}$ in (8.1) can now be simplified for potential flows. The continuity equation implies that the Laplacian of the potential function vanishes.

$$\begin{aligned}&\underline{\nabla} \cdot \underline{V} = 0 \Rightarrow \underline{\nabla} \cdot (\underline{\nabla}\phi) = 0, \text{ or}\\&\nabla^2\phi = 0,\end{aligned} \tag{8.33}$$

where the arguments $\underline{X}$ and t have been dropped for brevity.

Next we substitute $\underline{V}$ by $\underline{\nabla}\phi$ in the governing equation of $\underline{V}$ in (8.1)

$$\frac{\partial(\underline{\nabla}\phi)}{\partial t} + \left[(\underline{\nabla}\phi) \cdot \underline{\nabla}\right](\underline{\nabla}\phi) = -\frac{1}{\rho}\underline{\nabla}p + \underline{g}. \tag{8.34}$$

We can recast the advection term on the lhs of the last equation using a tensor identity that we derived earlier in Chap. 1 (1.70)

$$\left[(\underline{\nabla}\phi) \cdot \underline{\nabla}\right]\underline{\nabla}\phi = \frac{1}{2}\underline{\nabla}\left[(\underline{\nabla}\phi) \cdot (\underline{\nabla}\phi)\right]. \tag{8.35}$$

Using (8.35) in (8.34) we arrive at

$$\frac{\partial(\underline{\nabla}\phi)}{\partial t} + \frac{1}{2}\underline{\nabla}\left[(\underline{\nabla}\phi) \cdot (\underline{\nabla}\phi)\right] = -\frac{1}{\rho}\underline{\nabla}p + \underline{g}. \tag{8.36}$$

At this point we assume that $\underline{g} = -g\hat{e}_2$. This term can then be alternatively expressed as $-g\hat{e}_2 = -\underline{\nabla}(gx_2)$ leading to the following modified form of (8.36):

$$\underline{\nabla}\left[\frac{1}{g}\frac{\partial\phi}{\partial t} + \frac{(\underline{\nabla}\phi) \cdot (\underline{\nabla}\phi)}{2g} + \frac{p}{\rho g} + x_2\right] = 0, \tag{8.37}$$

which, in turn, implies that the entire argument of the gradient operator on the lhs of (8.37) is a constant in space. If A and B are any two locations in the flow field, then

$$\begin{aligned}&\left[\frac{1}{g}\frac{\partial\phi}{\partial t} + \frac{(\underline{\nabla}\phi) \cdot (\underline{\nabla}\phi)}{2g} + \frac{p}{\rho g} + x_2\right]_A = \\&\qquad\left[\frac{1}{g}\frac{\partial\phi}{\partial t} + \frac{(\underline{\nabla}\phi) \cdot (\underline{\nabla}\phi)}{2g} + \frac{p}{\rho g} + x_2\right]_B,\end{aligned} \tag{8.38}$$

where x_2 is the second coordinate of the respective location measured in the background Cartesian coordinate system. Further, we assume that the velocity field is steady, and thus, the potential function must not depend on time, leading to

$$\left[\frac{(\underline{\nabla}\phi)\cdot(\underline{\nabla}\phi)}{2g}+\frac{p}{\rho g}+x_2\right]_A=\left[\frac{(\underline{\nabla}\phi)\cdot(\underline{\nabla}\phi)}{2g}+\frac{p}{\rho g}+x_2\right]_B. \tag{8.39}$$

Alternatively, (8.39) can be expressed in terms of the magnitude of the velocity vector (V) by substituting $(\underline{\nabla}\phi)\cdot(\underline{\nabla}\phi)=V^2$,

$$\left[\frac{V^2}{2g}+\frac{p}{\rho g}+x_2\right]_A=\left[\frac{V^2}{2g}+\frac{p}{\rho g}+x_2\right]_B. \tag{8.40}$$

Indeed, Eq. (8.40) looks similar to Bernoulli's equation derived earlier in (8.21). Equation (8.40) is called the *generalized Bernoulli's equation*. The significance of the word *generalized* is that, unlike Bernoulli's equation, which requires the two chosen locations to be on the same streamline, for (8.40), the two chosen points (A and B) have no such restriction as long as they are within the domain of the same potential flow. While the restriction of the two chosen points being on the same streamline does not apply to the generalized Bernoulli's equation, it does require this "new" restriction that the entire flow domain must be vorticity-free. In contrast, Bernoulli's equation (8.21) is not subject to this condition. In summary, the generalized Bernoulli equation (8.40) has been derived under the following set of conditions: (i) the density of the fluid is constant, (ii) the fluid is ideal, (iii) the velocity field is steady, (iv) the body force is constant and acts along $-\hat{e}_2$ and (v) the flow field has $\underline{\omega}=0$ everywhere in the domain (and thus the velocity vector can be expressed as the gradient of a scalar field).

For a flow field with known far-upstream conditions (see Fig. 8.3), the location A can be conveniently chosen to be at ($x_1=-\infty$, $x_2=0$, x_3=0) as the reference point where both pressure $(p)_A$ and flow speed $(V)_A$ are known. Equation (8.40) can then be used to find pressure ($(p)_B$) at any other arbitrary location (B) in the flow domain. The velocity, at that arbitrary location, $(\underline{V})_B$ can be obtained by solving $\underline{V}=\underline{\nabla}\phi$. The potential function itself is to be determined by solving the transformed version of the continuity equation (8.33). Thus, in summary, the governing equation set for a steady potential flow is

$$\begin{aligned}
&\nabla^2\phi=0,\\
&(p)_B=(p)_A+\left(\frac{\rho V^2}{2}\right)_A-\left(\frac{\rho V^2}{2}\right)_B+\rho g\left[(x_2)_A-(x_2)_B\right],\\
&\underline{V}=\underline{\nabla}\phi.
\end{aligned} \tag{8.41}$$

While the continuity equation in set (8.41) is still a partial differential equation, the equation for pressure is an algebraic equation. The Laplacian equation is a linear equation. The linearity of the partial differential equation ($\nabla^2\phi=0$) implies that if two different known functions, ϕ_1 and ϕ_2, separately satisfy the continuity equation, then any linear combination of these two functions will also satisfy the continuity equation. This is called the *superposition principle* and is illustrated in (8.42) for two functions ϕ_1 and ϕ_2.

$$\text{If } \nabla^2\phi_1 = 0 \text{ and } \nabla^2\phi_2 = 0 \Rightarrow \nabla^2(\phi_3) = 0, \tag{8.42}$$

where $\phi_3 = \alpha\phi_1 + \beta\phi_2$ and the coefficients α and β are arbitrary constants. The superposition principle means that if we already know a set of independent scalar functions (ϕ's) which satisfies the equation $\nabla^2\phi = 0$, then we need not solve this partial differential equation for complex flow fields. Instead, we can propose the potential function of a complex flow field as a linear combination of known independent scalar functions along with a set of unknown constant coefficients. Subsequently, the boundary conditions of the flow domain can be leveraged to determine these unknown coefficients. Once these constant coefficients have been determined, we can find the velocity field as the gradient of the known combination potential function. Subsequently, the generalized Bernoulli equation (8.40) can be used to find pressure in the flow domain. We illustrate this solution procedure in detail for a potential flow field wherein the velocity field is known to be two-dimensional (2D) and has two scalar components (2C).

8.5 Potential Flows with 2C, 2D Velocity Fields

For a steady potential flow having a 2C, 2D velocity field, three simple and mutually independent functions are known to satisfy the continuity equation $\nabla^2\phi = 0$. The first of these functions expressed in the Cartesian coordinates (x_1, x_2) is

$$\phi\,(x_1, x_2) = V_o x_1 cos\alpha + V_o x_2 sin\alpha, \tag{8.43}$$

where α and V_o are constants. The two non-zero components of the velocity vector (V_1 and V_2) can be derived as

$$\begin{aligned} \underline{V} = \underline{\nabla}\phi \Rightarrow V_1\hat{e}_1 + V_2\hat{e}_2 &= \frac{\partial\phi}{\partial x_1}\hat{e}_1 + \frac{\partial\phi}{\partial x_2}\hat{e}_2, \\ \Rightarrow \underline{V}(x_1, x_2) &= V_o cos\alpha\hat{e}_1 + V_o sin\alpha\hat{e}_2. \end{aligned} \tag{8.44}$$

Equation (8.44) represents a uniform (constant in space) 2C, 2D velocity field. To visualize this flow field, in Fig. 8.4, we show representative streamlines in the plane of $x_1(\hat{e}_1) - x_2(\hat{e}_2)$. These streamlines are parallel and aligned to the direction of $\hat{e}_1$ at an angle α. Further, using the generalized Bernoulli's equation (8.40), pressure at an arbitrary point (B) in the flow field with A as the reference point is expressed as,

$$(p)_B = (p)_A + \rho g\,[(x_2)_A - (x_2)_B]\,, \tag{8.45}$$

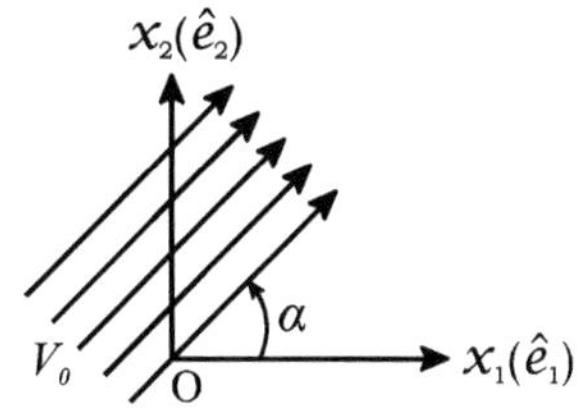

Fig. 8.4 Streamlines in a uniform velocity field

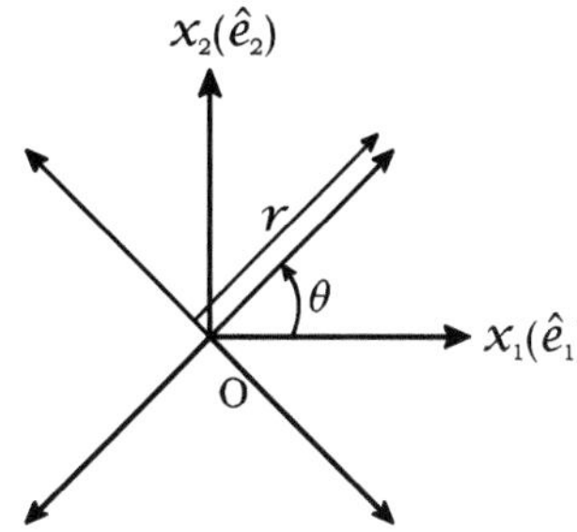

Fig. 8.5 Streamlines in a flow field with a source placed at the origin ($q > 0$)

which is identical to the hydrostatic pressure distribution derived in Chap. 7 (7.6).

The second simple solution of the continuity equation, when expressed using the cylindrical-polar coordinates (r, θ), is

$$\phi\,(r,\theta) = \frac{q}{2\pi} lnr, \tag{8.46}$$

where q is a constant. The two non-zero components of the velocity vector in the cylindrical-polar coordinate system (V_r and V_θ) can be derived as:

$$\begin{aligned} \underline{V} &= \underline{\nabla}\phi \Rightarrow V_r\hat{e}_r + V_\theta\hat{e}_\theta = \frac{\partial\phi}{\partial r}\hat{e}_r + \frac{1}{r}\frac{\partial\phi}{\partial\theta}\hat{e}_\theta, \\ &\Rightarrow \underline{V} = \frac{q}{2\pi r}\hat{e}_r. \end{aligned} \tag{8.47}$$

To visualize this flow field, in Fig. 8.5, we show representative streamlines in the plane of $x_1(\hat{e}_1) - x_2(\hat{e}_2)$ with a chosen positive value of q. These streamlines are along the radial direction $\hat{e}_r$. The streamlines appear as if a source of fluid is placed at the origin, and the fluid particles are emanating out in the radial direction. If $q < 0$, the streamlines point radially inwards toward the origin, as if there is a sink placed at the origin. Accordingly, the corresponding potential function (8.46) is called the potential function of a source (if $q > 0$) or the potential function of a sink (if $q < 0$). The quantity q is called the strength of the source/sink. Even though the direction of the velocity vector is oriented along $\hat{e}_r$ at all locations in the flow field, the magnitude of velocity still varies as $1/r$ (8.47). If A is a reference location at the location (r_A, θ_A) where the velocity and pressure are known, then the generalized Bernoulli's equation (8.40) can be employed to find the pressure at another location

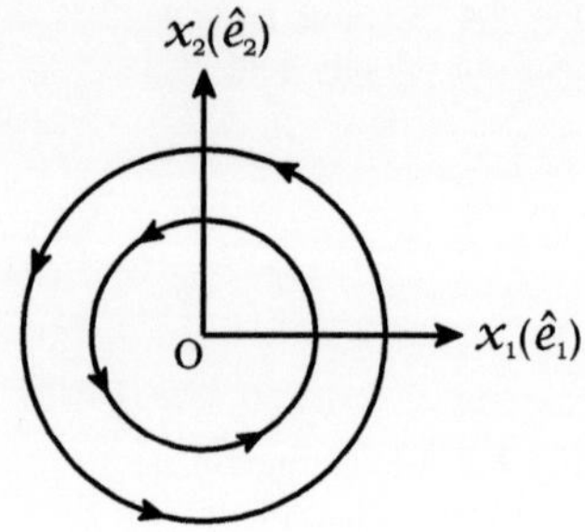

Fig. 8.6 Streamlines in a flow field with an anti-clockwise ($\Gamma > 0$) vortex having it center at the origin

B (r_B, θ_B)

$$(p)_B = (p)_A + \frac{\rho q^2}{8\pi^2}\left(\frac{1}{r_A^2} - \frac{1}{r_B^2}\right) + \rho g\left[(x_2)_A - (x_2)_B\right]. \tag{8.48}$$

The third simple solution of the continuity equation, when expressed using the cylindrical-polar coordinates (r, θ), is

$$\phi\,(r, \theta) = \frac{\Gamma\theta}{2\pi}, \tag{8.49}$$

where Γ is a constant. The two non-zero components of the velocity vector in the cylindrical-polar coordinate system can be derived as

$$\begin{aligned} \underline{V} &= \underline{\nabla}\phi \Rightarrow V_r\hat{e}_r + V_\theta\hat{e}_\theta = \frac{\partial\phi}{\partial r}\hat{e}_r + \frac{1}{r}\frac{\partial\phi}{\partial\theta}\hat{e}_\theta, \\ \Rightarrow \underline{V} &= \frac{\Gamma}{2\pi r}\hat{e}_\theta. \end{aligned} \tag{8.50}$$

To visualize this flow field, in Fig. 8.6, we show some representative streamlines in the plane of $x_1(\hat{e}_1) - x_2(\hat{e}_2)$ with a chosen positive value of Γ. These streamlines are along the tangential direction $\hat{e}_\theta$. If Γ is chosen to be a negative quantity, the streamlines would point along $-\hat{e}_\theta$ (clockwise direction). The corresponding potential function (8.49) is called the potential function of a *free vortex*. The quantity Γ is called the strength of the free vortex. Even though the direction of the velocity vector is oriented along $\hat{e}_\theta$ at all locations in the flow field of a free vortex, the magnitude of velocity still varies as $1/r$. If A is a reference location (r_A, θ_A) where the velocity and pressure are known, then the generalized Bernoulli's equation can (8.40) be employed to find the pressure at another location B (r_B, θ_B)

$$(p)_B = (p)_A + \frac{\rho\Gamma^2}{8\pi^2}\left(\frac{1}{r_A^2} - \frac{1}{r_B^2}\right) + \rho g\left[(x_2)_A - (x_2)_B\right]. \tag{8.51}$$

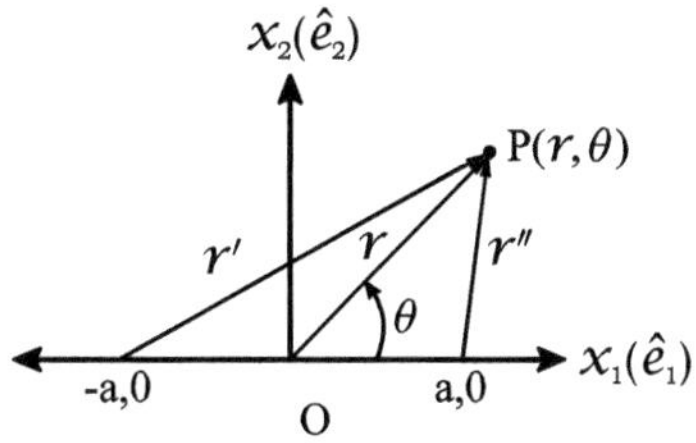

Fig. 8.7 Placement of a source and a sink for creating a doublet

The velocity fields of the source, the sink, and that of the free vortex field are not defined at the origin where $r = 0$. Such a point is called a *singular point* of the respective potential flow field. No solution (neither velocity nor pressure) can be inferred at such singular points.

Equations (8.48) and (8.51) imply that in the absence of a body force, the pressure would increase in the radially out direction in all three flow fields: that of a source, a sink, and a free vortex. In the flow field of the source, the radial gradient of pressure retards the motion of individual fluid particles as they move away from the origin. In contrast, in the flow field of a sink, the pressure gradient causes an increase in the speed of the particle as it moves toward the origin. In the case of a free vortex, the pressure gradient provides centripetal acceleration to the fluid particles so that each one of them continues to move on a circular path with a constant speed.

8.6 A Doublet

As mentioned in Sect. 8.4, any linear combination of three basic potential functions (8.43), (8.46), and (8.49) must also be a solution of the continuity equation (8.33). One such special combination of the potential functions of a source and that of a sink leads to a flow field called the flow field of a *doublet*. The utility of a doublet is that later, it can be used in combination with the potential function of a uniform stream to generate the solution for flows past some objects of simple geometrical shapes.

To create a doublet, we consider a source and a sink placed close to each other such that the origin of the source is at the location (–a, 0) and that of the sink is at the location (a, 0) (referring to a frame-fixed Cartesian coordinate system, Fig. 8.7). The strength of the source is q (>0) and that of the sink is $-q$.

At an arbitrary location P $(r(\neq 0), \theta)$ the potential function of the source is

$$\phi_{source}(r, \theta) = \frac{q}{2\pi} lnr', \tag{8.52}$$

and that of the sink is

$$\phi_{sink}(r, \theta) = -\frac{q}{2\pi} lnr''. \tag{8.53}$$

The quantities r' and r'' denote the distances of the location P from the locations where the source and the sink are placed. These quantities can be expressed in terms of a, r and θ as (Fig. 8.7)

$$r'' = \left[(r sin\theta)^2 + (r cos\theta - a)^2\right]^{1/2} = r\left[1 + \left(\frac{a}{r}\right)^2 - \frac{2a}{r}cos\theta\right]^{1/2},$$
$$r' = \left[(r sin\theta)^2 + (r cos\theta + a)^2\right]^{1/2} = r\left[1 + \left(\frac{a}{r}\right)^2 + \frac{2a}{r}cos\theta\right]^{1/2}. \quad (8.54)$$

We consider the potential function which is the sum of ϕ_{source} and ϕ_{sink}

$$\phi(r,\theta) = \phi_{source}(r,\theta) + \phi_{sink}(r,\theta), \text{ or} \quad (8.55)$$
$$\phi(r,\theta) = \frac{q}{2\pi}ln r' - \frac{q}{2\pi}ln r''.$$

Using (8.54) in (8.55), $\phi(r,\theta)$ is expressed as

$$\phi(r,\theta) = \frac{q}{4\pi}\left\{ln\left[1 + \left(\frac{a^2}{r^2} + \frac{2a}{r}cos\theta\right)\right] - ln\left[1 + \left(\frac{a^2}{r^2} - \frac{2a}{r}cos\theta\right)\right]\right\}. \quad (8.56)$$

At this point we assume (which is a part of the definition of a doublet) that

$$\left|\frac{a^2}{r^2} \pm \frac{2a}{r}cos\theta\right| << 1, \quad (8.57)$$

where $|.|$ denotes the magnitude of the quantity of interest. Accordingly, the two terms involving ln are to be first expressed as a series expansion and subsequently truncated up to the first order

$$\begin{aligned} ln\left[1 + \left(\frac{a^2}{r^2} + \frac{2a}{r}cos\theta\right)\right] &= \left(\frac{a^2}{r^2} + \frac{2a}{r}cos\theta\right) - \frac{1}{2}\left(\frac{a^2}{r^2} + \frac{2a}{r}cos\theta\right)^2 \\ &+ \frac{1}{3}\left(\frac{a^2}{r^2} + \frac{2a}{r}cos\theta\right)^3 + H.O.T., \\ &\approx \left(\frac{a^2}{r^2} + \frac{2a}{r}cos\theta\right), \end{aligned}$$

$$\begin{aligned} ln\left[1 + \left(\frac{a^2}{r^2} - \frac{2a}{r}cos\theta\right)\right] &= \left(\frac{a^2}{r^2} - \frac{2a}{r}cos\theta\right) - \frac{1}{2}\left(\frac{a^2}{r^2} - \frac{2a}{r}cos\theta\right)^2 \\ &+ \frac{1}{3}\left(\frac{a^2}{r^2} - \frac{2a}{r}cos\theta\right)^3 + H.O.T., \\ &\approx \left(\frac{a^2}{r^2} - \frac{2a}{r}cos\theta\right). \end{aligned} \quad (8.58)$$

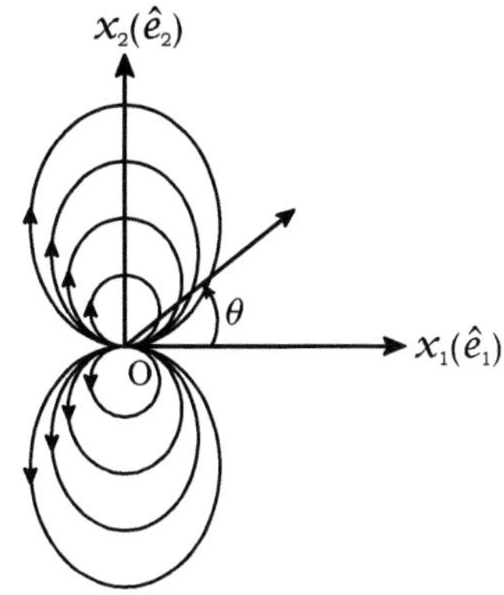

Fig. 8.8 Streamlines in the flow field of a doublet

Using (8.58) in (8.56) leads to the following expression:

$$\phi(r,\theta) \approx \frac{q}{4\pi}\left\{\frac{2acos\theta}{r} - \left(-\frac{2acos\theta}{r}\right)\right\} = \frac{aqcos\theta}{\pi r}. \tag{8.59}$$

Since the quantities a and q appear together as a product in the rhs of (8.59), and we have already assumed a to be an arbitrarily small quantity, we choose (a part of the definition of a doublet) q such that even though a is small, the product aq is a desired number. We replace the product aq with a new quantity K (>0), which is called the strength of the doublet. In the limit, $a \to 0$ while $aq \to K$ (8.59) leads to the potential function of a doublet

$$\phi_{doublet} = \lim_{a\to 0,\ aq\to K} \frac{aqcos\theta}{\pi r} = \frac{Kcos\theta}{\pi r}. \tag{8.60}$$

Like the potential functions of a source and a sink, neither velocity nor pressure is defined at $r = 0$ (a singular point) in (8.60).

The two non-zero components of the velocity vector in the potential flow field of a doublet in the cylindrical-polar coordinate system (V_r and V_θ) are derived as

$$\begin{aligned} \underline{V} = \underline{\nabla}\phi &\Rightarrow V_r\hat{e}_r + V_\theta\hat{e}_\theta = \frac{\partial\phi}{\partial r}\hat{e}_r + \frac{1}{r}\frac{\partial\phi}{\partial\theta}\hat{e}_\theta, \\ \Rightarrow \underline{V} &= -\frac{Kcos\theta}{\pi r^2}\hat{e}_r - \frac{Ksin\theta}{\pi r^2}\hat{e}_\theta. \end{aligned} \tag{8.61}$$

To visualize this flow field, in Fig. 8.8, we show some representative streamlines in the plane of $x_1(\hat{e}_1) - x_2(\hat{e}_2)$ with a chosen value of K.

If A is a reference location (r_A, θ_A) where the velocity and pressure are known, then the generalized Bernoulli's equation (8.40) can be employed to find the pressure at another location B (r_B, θ_B)

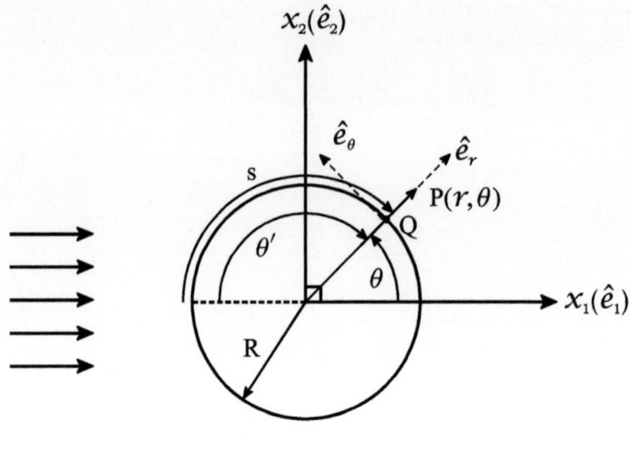

Fig. 8.9 A long cylinder of radius R exposed to a uniform stream

$$(p)_B = (p)_A + \frac{\rho K^2}{2\pi^2}\left(\frac{1}{r_A^4} - \frac{1}{r_B^4}\right) + \rho g\left[(x_2)_A - (x_2)_B\right]. \tag{8.62}$$

8.7 Solution of Potential Flow Past a Long Circular Cylinder

In this section, we demonstrate how the superposition of the potential function of a uniform flow with the potential function of a doublet can lead to the solution of the potential flow that exists past a solid cylinder of radius R. The cylinder is fixed to the inertial reference frame in context. It is exposed to a velocity field which is uniform in the far-upstream region (Fig. 8.9). The central axis of the cylinder coincides with the $x_3(\hat{e}_3)$ axis. The cylinder is assumed to have an infinite length along this axis such that the associated potential flow field has a 2C, 2D velocity field. We wish to find $\underline{V}(r, \theta)$ and $p(r, \theta)$.

We consider the superposition of the potential functions of a doublet (with its singular point placed at the origin of the coordinate system) and that of a known uniform stream, $Vo\hat{e}_1$, which is parallel to $\hat{e}_1$. Thus, at an arbitrary location P(r, θ) the combined potential function is

$$\phi(r, \theta) = \phi_{uniform\ flow}(r, \theta) + \phi_{doublet}(r, \theta), \tag{8.63}$$

where $\phi_{uniform\ flow}(r, \theta)$ and $\phi_{doublet}(r, \theta)$ are the potential functions of a uniform flow (8.43) and that of a doublet (8.60), respectively. The potential function of a uniform flow has earlier been presented using the Cartesian coordinate system in (8.43). In the current scenario $\alpha = 0$ (uniform flow streamlines are along the $\hat{e}_1$ direction). Further, we convert x_1 into the cylindrical-polar coordinate system by substituting $x_1 = rcos\theta$, Thus,

$$\phi_{uniformflow}(r, \theta) = V_o r cos\theta. \tag{8.64}$$

Using (8.64) and (8.60) in (8.63) we arrive at

$$\phi(r, \theta) = V_o r cos\theta + \frac{K cos\theta}{\pi r}. \tag{8.65}$$

At this point, even though V_o is known (the far-upstream condition), K is an unknown constant. Taking the gradient of the combined potential function (8.65) we arrive at the velocity vector

$$\underline{V} = \underline{\nabla}\phi \Rightarrow V_r \hat{e}_r + V_\theta \hat{e}_\theta = \frac{\partial \phi}{\partial r}\hat{e}_r + \frac{1}{r}\frac{\partial \phi}{\partial \theta}\hat{e}_\theta,$$
$$\Rightarrow \underline{V} = \left(V_o cos\theta - \frac{K cos\theta}{\pi r^2}\right)\hat{e}_r + \left(-V_o sin\theta - \frac{K sin\theta}{\pi r^2}\right)\hat{e}_\theta. \tag{8.66}$$

For this velocity field to represent the flow past a cylinder of radius R, it must satisfy the no-penetration condition at $r = R$ at all $\theta \in [0, 2\pi]$. The no-penetration condition on the surface of the cylinder requires

$$\underline{V} \cdot \hat{e}_n = 0 \text{ at } r = R \text{ for all } \theta \in [0, 2\pi]. \tag{8.67}$$

where $\hat{e_n}$ is the local outward normal on the surface of the solid cylinder. Since the origin of our working coordinate system coincides with the center of the cylinder (Fig. 8.9), (8.67) implies that the radial component of the velocity vector on the surface of the solid cylinder must vanish to satisfy the no-penetration condition.

$$V_r = 0 \text{ at } r = R \text{ for all } \theta \in [0, 2\pi] \Rightarrow V_o cos\theta - \frac{K cos\theta}{\pi R^2} = 0. \tag{8.68}$$

Equation (8.68) is satisfied at all θ, if $K = V_o \pi R^2$. Having determined K, we can now use this expression of K in (8.66) to find the velocity field $\underline{V}(r, \theta)$:

$$\underline{V} = V_o cos\theta \left(1 - \frac{R^2}{r^2}\right)\hat{e}_r - V_o sin\theta \left(1 + \frac{R^2}{r^2}\right)\hat{e}_\theta. \tag{8.69}$$

Although (8.69) is mathematically valid at all locations with $r > 0$, the velocity field solution at locations with $r < R$ is not relevant. We are interested in finding the velocity field only over the solid cylinder of radius R. Thus, the mathematical solution at $r < R$ is ignored. At all other locations with $r \geq R$ and $0 \leq \theta \leq 2\pi$, Eq. (8.69) does describe the velocity field of the potential flow that exists past the cylinder of radius R. Based on this velocity field, Fig. 8.10 shows some representative streamlines in the flow field.

To find the pressure field, we appeal to the generalized Bernoulli's equation (8.40). We take a reference point A located at ($x_1 = -\infty$, $x_2 = 0$, x_3=0). At that location,

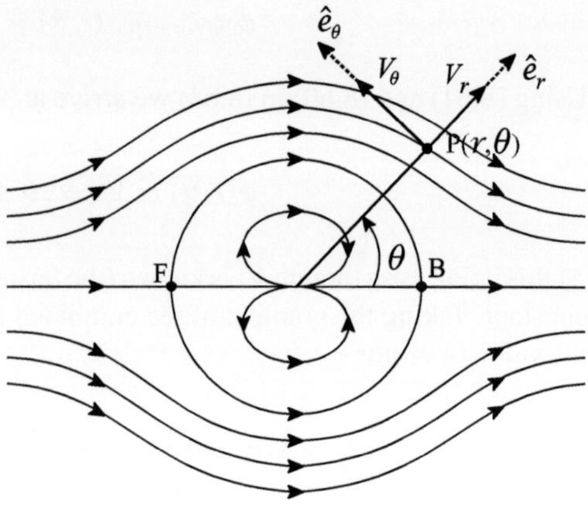

Fig. 8.10 Streamlines of the potential flow past a cylinder of radius R

the velocity vector and pressure are both known ($V_o\hat{e}_1$, p_o). Thus, the pressure at any arbitrary location ($r \geq R$,θ) is described by the following equation:

$$p(r,\theta) = p_o + \frac{\rho}{2}\left(V_o^2 - V^2\right) - \rho g r sin\theta, \tag{8.70}$$

where V represents the speed of the fluid particle at (r, θ). Using (8.69), we find

$$V^2 = V_o^2\left[1 + \left(\frac{R}{r}\right)^4 + 2\left(\frac{R}{r}\right)^2\left(2sin^2\theta - 1\right)\right]. \tag{8.71}$$

Thus, (8.70) is modified to

$$p(r,\theta) = p_o - \frac{\rho V_o^2}{2}\left[\left(\frac{R}{r}\right)^4 + 2\left(\frac{R}{r}\right)^2\left(2sin^2\theta - 1\right)\right] - \rho g r sin\theta. \tag{8.72}$$

In fluid mechanics, the quantity $\rho V^2/2$ is referred to as the *local dynamic pressure* of the flow field. Thus, the quantity $\rho V_o^2/2$ is the dynamic pressure of the free stream (Fig. 8.9). In (8.72) we observe that the pressure field takes three contributions. The first contribution is p_o, and it represents the influence of the boundary pressure (from the far upstream) on the local pressure at (r,θ). The third contribution is the influence of the body force on the local pressure. The middle term represents the influence of the velocity field on pressure. Therein, two factors can be identified. The first factor is $\rho V_o^2/2$, which is the dynamic pressure of the free stream. The second factor $\left[\left(\frac{R}{r}\right)^4 + 2\left(\frac{R}{r}\right)^2\left(2sin^2\theta - 1\right)\right]$ is a non-dimensional quantity that depends on (i) current location (r,θ) and (ii) the geometry of the cylinder (the blockage offered by the solid object placed in an otherwise uniform flow).

At locations which are far away from the surface of the cylinder ($r \to \infty$), (8.69) shows that the velocity vector tends to become $V_o\hat{e}_1$, which is the same as the upstream velocity.

$$\lim_{r\to\infty} \underline{V}(r,\theta) = \lim_{r\to\infty} \left[V_o cos\theta \left(1 - \frac{R^2}{r^2}\right)\hat{e}_r - V_o sin\theta \left(1 + \frac{R^2}{r^2}\right)\hat{e}_\theta \right],$$
$$= V_o cos\theta \hat{e}_r - V_o sin\theta \hat{e}_\theta = V_o \hat{e}_1. \quad (8.73)$$

Further, As $r \to \infty$, the limiting expression for the pressure (8.72) becomes purely hydrostatic

$$\lim_{r\to\infty} p(r,\theta) = \lim_{r\to\infty} \left\{ p_o - \frac{\rho V_o^2}{2} \left[\left(\frac{R}{r}\right)^4 + 2\left(\frac{R}{r}\right)^2 (2sin^2\theta - 1) \right] \right.$$
$$\left. -\rho g r sin\theta \right\},$$
$$= p_o - \rho g x_2. \quad (8.74)$$

When the cylinder is placed in a flow field which is otherwise uniform, the no-penetration condition imposed by the surface of the cylinder (or the physical blockage offered by the cylinder) forces the velocity field to change in accordance with the continuity equation. This modified velocity field, in accordance with the generalized Bernoulli's equation, forces the pressure field to depart from the otherwise purely hydrostatic field pressure distribution that would exist in a uniform flow. This influence of the blockage is most evident near the cylinder ($r \to R$) and tends to vanish as we move to faraway locations ($r \to \infty$) in a potential flow field.

Of special interest is to examine the variations in velocity and pressure right on the cylinder surface ($r = R$). Owing to the imposed no-penetration condition, the velocity on the cylinder surface is purely tangential to the surface (setting $r = R$ in Eq. 8.71).

$$\underline{V} = -2V_o sin\theta \hat{e}_\theta. \quad (8.75)$$

The negative sign implies that the tangential velocity is opposite to the local tangential vector, which points in the counter-clockwise direction (Fig. 8.9). Thus, the fluid particles that touch the cylinder's surface move in a clockwise direction. The pressure variation is expressed as (setting $r = R$ in Eq. 8.72)

$$p(r,\theta) = p_o + \frac{\rho V_o^2}{2}\left[1 - 4sin^2\theta\right] - \rho g r sin\theta. \quad (8.76)$$

In Fig. 8.11, we plot the variation in the magnitude of the velocity vector (V). This variation is presented as a function of θ' which equals $\pi - \theta$ (Fig. 8.9). With Eq. (8.69), it can be verified that the variation of the velocity magnitude is symmetric about the x_1 $(\hat{e}_1)$ axis. Thus, the variation in V in Fig. 8.11 is presented only in the range of $0 \leq \theta' \leq \pi$. We observe that the variation is non-monotonous. There are two points on the cylinder surface where the velocity completely vanishes. These points are at $\theta' = 0$ and $\theta' = \pi$. In general, such points in a flow field where $\underline{V}$ vanishes

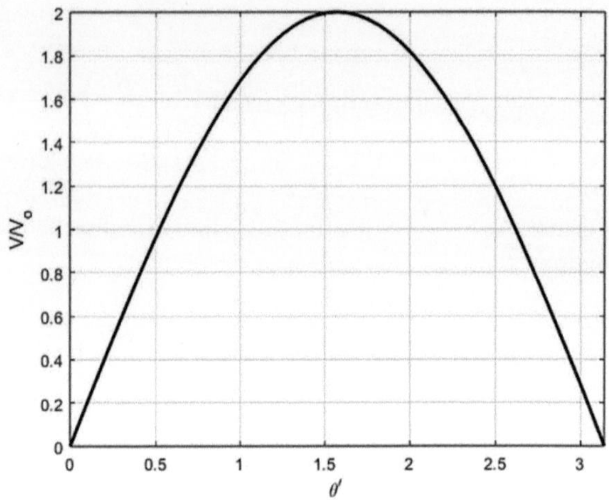

Fig. 8.11 Variation in the magnitude of the normalized velocity vector (V/V_o) on the cylinder surface versus θ' (radians)

are called the *stagnation points* of the flow field. Thus, the potential flow past a circular cylinder has two stagnation points. The stagnation point at ($r = R, \theta' = 0$) is called the forward stagnation point, whereas the stagnation point at ($r = R, \theta' = \pi$) is called the rear stagnation point. In between these two stagnation points, the variation in the magnitude of the velocity vector is sinusoidal, with the peak magnitude occurring at $\theta' = \pi/2$. Over the forward part of the cylinder ($0 < \theta' \leq \pi/2$), a fluid particle which is moving on the surface of the cylinder accelerates and reaches the maximum speed ($2V_o$) at $\theta' = \pi/2$. Subsequently, the fluid particle decelerates over the rear part of the cylinder ($\pi/2 \leq \theta' < \pi$).

Equation (8.76) shows that pressure at both the stagnation points ($\theta' = 0$ and $\theta' = \pi$) is $p_o + \frac{1}{2}\rho V_o^2$. In the fluid mechanics of constant-density fluids, we call the sum of local pressure (p) and the local dynamic pressure ($\frac{\rho}{2}V^2$) as the *local stagnation pressure* ($p + \frac{\rho V^2}{2}$). Since the velocity at a stagnation point vanishes, the local stagnation pressure at a stagnation point is identical to the local pressure there. However, the quantity $p_o + \frac{1}{2}\rho V_o^2$ is the stagnation pressure of the reference location that we chose in the far-upstream region ($x_1 = -\infty, x_2 = 0, x_3 = 0$). Thus, at the two stagnation points, the local pressure equals the stagnation pressure of the reference location.

The pressure variation on solid surfaces is often examined in terms of a non-dimensional quantity called the *coefficient of pressure*, and is represented by the symbol C_p. This quantity is defined as:

$$C_p(r, \theta) = \frac{p(r, \theta) - p_o}{\frac{\rho V_o^2}{2}}, \tag{8.77}$$

where p_o and V_o are the reference values of pressure and the magnitude of fluid velocity. The primary advantage of examining the variation in pressure in terms of

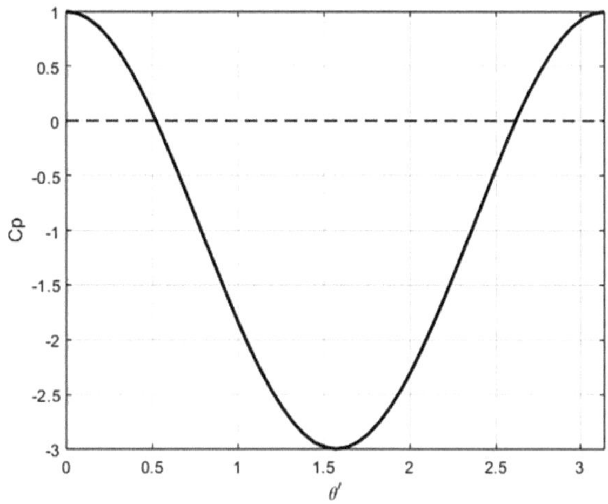

Fig. 8.12 Variation in C_p on the cylinder surface versus θ' (radians). Body forces are assumed to be zero for this illustration

C_p, rather than the absolute values of p, is that with appropriate choices of these reference values (p_o and V_o), C_p varies over a small range of numbers, and that too, being close to unity. On the other hand, in many engineering applications, the absolute value of pressure tends to be quite large (atmospheric pressure at standard sea level conditions is $1.01 \times 10^5 Nm^{-2}$). Another advantage of using C_p is that it focuses on the variation of local pressure relative to the reference pressure (p_o). Indeed, as it is clear in (8.76), the influence of the no-penetration condition on the local pressure, and for that matter, even the influence of body force on the local pressure, is to shift the pressure value above or below the reference level (p_o) (Fig. 8.12).

To gain further insight into specifically how the no-penetration condition influences the pressure variation on the cylinder surface, we ignore the influence of the body force and focus on the remaining expression of (8.76). Thus, the variation in C_p over the cylinder surface is

$$C_p = 1 - 4sin^2\theta = 1 - 4sin^2\theta^{'}. \tag{8.78}$$

This variation is graphically presented in (Fig. 8.12). We observe that at the two stagnation points ($\theta^{'} = 0$ and $\theta^{'} = \pi$) C_p equals unity. Between the two stagnation points, on the forward part of the cylinder, C_p decreases and reaches its minimum value (-3) at $\theta^{'} = \pi/2$. Further, downstream along the cylinder surface, C_p increases and reaches unity at the rear stagnation point ($\theta^{'} = \pi$). This trend is the exact opposite of that identified in Fig. 8.11. Indeed, the location on the cylinder surface where the magnitude of the velocity peaks is identical to the location where the C_p ($p(r, \theta)$) attains its minimum value. This location divides the cylinder surface into two parts. The part which is upstream of the location and where the gradient of pressure along the body surface, $\frac{\partial p}{\partial s}$, (where s is the distance of a location on the surface measured from the forward stagnation point, see Fig. 8.9) is negative is called the region of *favorable pressure gradient*. In this region, the net pressure force acting on a fluid particle acts in the same direction as the velocity vector of the fluid particle. On the other hand, on the part of the surface which is downstream of the location where the C_p attains its minimum value, the gradient of pressure along the body surface $(\frac{\partial p}{\partial s})$ is positive is called the region of *adverse pressure gradient.* On this part of the cylinder surface, a fluid particle moving on the surface of the cylinder experiences a net pressure force, which is directed opposite to its own velocity vector.

8.8 Solution of Potential Flow Past Bodies of Arbitrary Shapes

The procedure of using a combination of simple known solutions of $\nabla^2\phi = 0$ can be extended to simulate flow fields past bodies of arbitrary shapes, too. In this extended procedure, a large but finite number of elementary solutions (sources/sinks/free vortices/doublet) with unknown coefficients (q's, Γ's or K's) are distributed in space. The algebraic expression of the resulting potential function is derived using the superposition principle. Subsequently, the no-penetration boundary condition is imposed at a number of discrete locations where a solid surface exists. The number of such discrete points must be the same as the number of unknown coefficients under consideration. This process results in a system of algebraic equations involving unknown coefficients. Numerically solving this algebraic equation set determines the unknown coefficients with which the combined potential function of the flow field can be computed. Subsequently, the velocity components are extracted from this computed potential field using (8.32). Finally, using a reference point where the local speed and the local pressure of the flow are known, along with the generalized Bernoulli's equation (8.40), the pressure field is computed. Such methods are called the *panel methods* of seeking solutions of potential flow fields. Any further detailed discussion of such methods is outside the scope of the current text (see [3] for more details). Even though some numerical algorithms, computer programming, and subsequent computer processing are required to obtain the velocity and the pressure field using the panel methods, the resources and effort required are still, typically, much less than what it would take to numerically solve the governing partial differential equation (8.1).

Chapter 9
Some Simple Representative Flows of a Viscous Fluid

Flows of real fluids (also called the *V!viscous fluids*) are categorized as being *laminar*, *transitional flows*, or *turbulent*. Turbulent flows are always unsteady, and the velocity field therein is always 3C and 3D. Further, turbulent flows are always associated with non-zero vorticity. In contrast, laminar flows can be steady or unsteady, and their velocity field can be 3C, 2C, or even 1C. Transitional flows, too, are always unsteady. The onset of transnational regime coincides with the demise of the laminar nature of a flow field. Without delving into any advanced topics explaining why and how a flow field may behave as transitional/turbulent, we wish to focus only on laminar flow fields in this book. Again, the author would like to remind the reader that, in this chapter, our aim is to consider some special viscous flow fields wherein the Navier–Stokes equation set of a constant density flow field (5.110) can be reasonably simplified to the extent that we can solve them without the need to have any advanced CFD technique. The solution obtained thereupon can be used to develop some insights into the velocity and pressure fields of a real fluid. Specifically, there are two such flow fields examined in this chapter: steady laminar flow over a flat plate, and steady laminar flow between two parallel plates.

9.1 Flat Plate Boundary Layer

In Fig. 9.1, we show a thin solid flat plate lying in the $x_1\left(\hat{e}_1\right) - x_3\left(\hat{e}_3\right)$ plane of the frame-fixed Cartesian coordinate system $Ox_1\left(\hat{e}_1\right) - x_2\left(\hat{e}_2\right)x_3\left(\hat{e}_3\right)$. The origin of the coordinate system coincides with the left edge of the plate, as shown in the figure. The plate is fixed with respect to the ground (an inertial reference frame). The plate is assumed to be extending indefinitely in the $\hat{e}_1$ direction and $\pm\hat{e}_3$ directions. In the far-upstream region, the fluid has a uniform velocity $V_o\hat{e}_1$ with respect to the ground frame.

© The Author(s), under exclusive license to Springer Nature Switzerland AG 2026

S. S. Sinha, *Fundamentals of Fluid Mechanics*,
https://doi.org/10.1007/978-3-032-04210-1_9

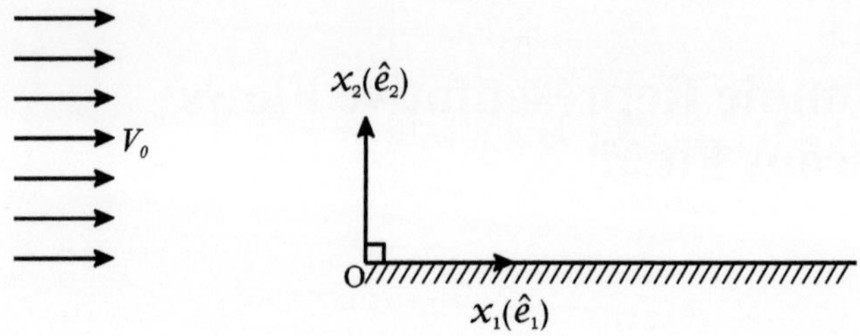

Fig. 9.1 A flat plate placed in an otherwise uniform stream

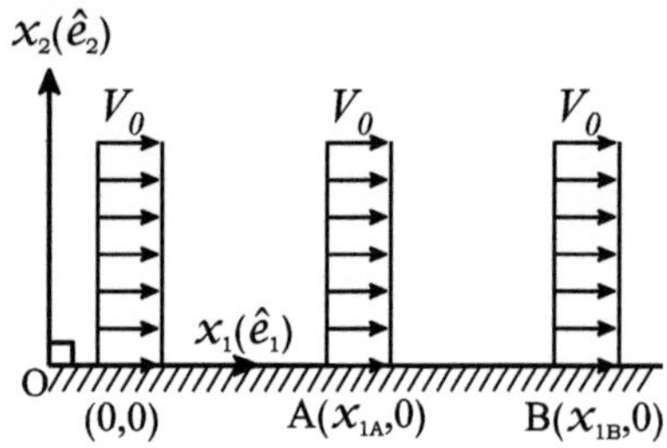

Fig. 9.2 V_1 versus x_2 profiles at different stream-wise (x_1) stations in an ideal fluid flow over a flat plate

Figure 9.2 shows the variation in V_1 versus x_2 at three representative x_1 stations for the case when $\mu = 0$. Such a plot is also called the *profile* of the *stream-wise* velocity component versus the *wall-normal* direction. These profiles must all be uniform as well as identical to each other. However, when the fluid is viscous, the plate imposes the no-penetration as well as the no-slip condition on the velocity of fluid particles that are in touch with the plate surface. Due to the no-slip condition $V_1(x_1, x_2 = 0) = 0$ at all x_1 stations along the plate. However, we expect that at locations far away from the wall, the influence of the no-slip condition imposed by the wall on the velocity field may diminish and consequently $V_1(x_1, x_2 \to \infty) \to V_o$. For the viscous flow field over a flat plate as shown in Fig. 9.1, at a chosen x_1 station, the wall-normal distance where the stream-wise velocity component (V_1) equals 99% of the free stream velocity is called the *boundary layer thickness* at that x_1 station.

$$\delta(x_1) = x_2 \text{ such that } V_1(x_1, x_2) = 0.99V_o. \tag{9.1}$$

The boundary layer thickness varies with x_1. The region of the flow field, which is bounded by the flat plate on one side and the curve $\delta(x_1)$ on the other side, is called the *boundary layer region* (Fig. 9.3). This is the region where the influence of the viscous forces is deemed to have a significant influence on the velocity field. Outside the boundary layer region, even though it is the same viscous fluid with an identical coefficient of viscosity (μ) as it is inside the boundary layer, the viscous forces have an insignificant influence on the motion of fluid particles. This may happen because outside the boundary layer, despite μ being non-zero, the components of strain-

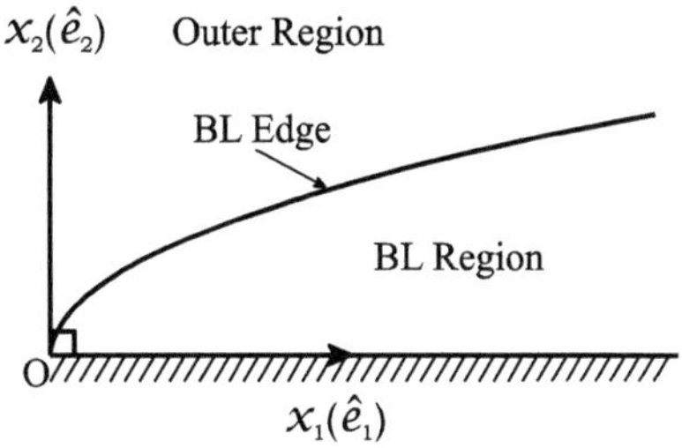

Fig. 9.3 Two regions of the flow field past a flat plate divided by the edge of the boundary layer (BL)

rate tensor may not be significant enough to cause the generation of any significant viscous forces (5.95). In Fig. 9.3, we refer to this region as the *outer region*.

The exact governing equations of the steady flow field over the flat plate of Fig. 9.1 are (5.110)

$$\underline{\nabla} \cdot \underline{V} = 0,$$
$$\rho \left(\underline{V} \cdot \underline{\nabla} \right) \underline{V} = -\underline{\nabla} p + \rho \underline{g} + \mu \nabla^2 \underline{V}. \tag{9.2}$$

We assume the flat plate to be infinitely long in the $\pm\hat{e}_3$ direction such that the laminar velocity field existing over the plate is 2C ($V_3 = 0$) and 2D (velocity field depends only on x_1 and x_2 alone). Further, we assume that the flow field is steady. In the Cartesian coordinate system (shown in Fig. 9.1), the continuity equation simplifies to

$$\frac{\partial V_1}{\partial x_1} + \frac{\partial V_2}{\partial x_2} = 0. \tag{9.3}$$

The governing equation of V_1 simplifies to

$$V_1 \frac{\partial V_1}{\partial x_1} + V_2 \frac{\partial V_1}{\partial x_2} = -\frac{1}{\rho} \frac{\partial p}{\partial x_1} + \nu \frac{\partial^2 V_1}{\partial x_1 \partial x_1} + \nu \frac{\partial^2 V_1}{\partial x_2 \partial x_2} + g_1. \tag{9.4}$$

The governing equation of V_2 simplifies to

$$V_1 \frac{\partial V_2}{\partial x_1} + V_2 \frac{\partial V_2}{\partial x_2} = -\frac{1}{\rho} \frac{\partial p}{\partial x_2} + \nu \frac{\partial^2 V_2}{\partial x_1 \partial x_1} + \nu \frac{\partial^2 V_2}{\partial x_2 \partial x_2} + g_2. \tag{9.5}$$

The governing equation of V_3 simplifies to

$$0 = -\frac{1}{\rho} \frac{\partial p}{\partial x_3}. \tag{9.6}$$

Here we have assumed that the vector $\underline{g}$ is oriented such $\underline{g} = g_1\hat{e}_1 + g_2\hat{e}_2$. Equation (9.6) implies that the pressure field must be a function of only x_1 and x_2. With this understanding, there is no need to explicitly list (9.6) anymore. The symbol ν appearing in (9.5) is called the *coefficient of kinematic viscosity* and equals μ/ρ.

9.1.1 Order-of-Magnitude Analysis

To further simplify this set of equations governing the fluid motion (9.3), (9.4), and (9.5), we analyze these equations following a methodology which is called the *order-of-magnitude* analysis. This procedure involves

1. Identifying some characteristic values of various variables.
2. Normalizing an equation using these characteristic values.
3. Estimating the order of magnitude of various terms in that equation.
4. Discarding the terms which are negligible compared to other terms.

This process may finally yield an approximate but simplified set of governing equations.

For our analysis in this section, we first adopt for ourselves a formal definition of the order of magnitude (OM) of a number (ζ). We say that the OM of ζ is 10^M (or $\mathcal{O}(\zeta) = 10^M$) if

$$10^{M-0.5} \leq |\zeta| < 10^{M+0.5}, \tag{9.7}$$

where M is an integer. For example, consider $\zeta = 1.5$. Since $10^{0-0.5} \leq 1.5 < 10^{0+0.5}$, following our definition (9.7), we conclude $\mathcal{O}(1.5) = 10^0 = 1$. If $\zeta = 15$, we find that $10^{1-0.5} \leq 15 < 10^{1+0.5}$, and we conclude that $\mathcal{O}(15) = 10^1 = 10$.

Knowledge of the order of magnitude of a variable in an equation (even if we do not know the exact value of the variable yet) may help us simplify an equation involving that variable. For example, consider a non-linear algebraic equation involving two unknown quantities: α and β such that

$$\alpha + \beta + \beta^2 = 29. \tag{9.8}$$

Suppose we have been informed that $\mathcal{O}(\beta) >> \mathcal{O}(\beta^2)$, then it is reasonable to approximate (9.8) to

$$\alpha + \beta \approx 29. \tag{9.9}$$

In this example, the resulting equation has now become simpler (in this case, the original non-linear equation in β has now become linear in β), which, in turn, may make the solution procedure easier.

Closely related to the idea of the OM of a number is the characteristic value of an Eulerian field variable, say $\phi(x_1, x_2)$. We say that ϕ_C is the characteristic value of the variable $\phi(x_1, x_2)$ over a domain of interest, if

$$\mathcal{O}\left[\frac{\phi(x_1, x_2)}{\phi_C}\right] = 10^0 = 1, \tag{9.10}$$

over most (if not all) of the domain of interest. The quantity ϕ_C is a chosen number such that (9.10) is satisfied over most of the domain of interest. Using ϕ_C, we now define a normalized version of the variable $\phi^*(x_1, x_2)$

$$\phi^*(x_1, x_2) = \frac{\phi(x_1, x_2)}{\phi_C}. \tag{9.11}$$

Equation (9.10) implies that if ϕ_C has been chosen aptly then

$$\mathcal{O}\left[\phi^*(x_1, x_2)\right] = 1, \tag{9.12}$$

over most of the domain of interest.

In the context of the governing equations of the flow over a flat plate, if we appropriately choose two numbers V_{1C} and V_{2C} as the characteristic values of the variables $V_1(x_1, x_2)$ and $V_2(x_1, x_2)$, then we can define the normalized version of these variables as

$$V_1^*(x_1, x_2) = \frac{V_1(x_1, x_2)}{V_{1C}} \text{ and } V_2^*(x_1, x_2) = \frac{V_2(x_1, x_2)}{V_{2C}}, \tag{9.13}$$

such that

$$\mathcal{O}\left[V_1^*(x_1, x_2)\right] = 1 \text{ and } \mathcal{O}\left[V_2^*(x_1, x_2)\right] = 1, \tag{9.14}$$

over most of the domain of interest.

Further, we define *characteristic length scales*, L_{1C} and L_{2C}, for a flow field. Using these characteristic length scales, the gradients of various scalar variables along $\hat{e}_1$ and $\hat{e}_2$ directions are to be normalized as

$$\begin{aligned} \left(\frac{\partial V_1}{\partial x_1}\right)^* &= \left(\frac{\partial V_1}{\partial x_1}\right)\frac{L_{1C}}{V_{1C}}, \left(\frac{\partial V_1}{\partial x_2}\right)^* = \left(\frac{\partial V_1}{\partial x_2}\right)\frac{L_{2C}}{V_{1C}}, \\ \left(\frac{\partial V_2}{\partial x_1}\right)^* &= \left(\frac{\partial V_2}{\partial x_1}\right)\frac{L_{1C}}{V_{2C}}, \left(\frac{\partial V_2}{\partial x_2}\right)^* = \left(\frac{\partial V_2}{\partial x_2}\right)\frac{L_{2C}}{V_{2C}}, \end{aligned} \tag{9.15}$$

such that

$$\begin{aligned} \mathcal{O}\left[\left(\frac{\partial V_1}{\partial x_1}\right)^*\right] &= 1, \mathcal{O}\left[\left(\frac{\partial V_1}{\partial x_2}\right)^*\right] = 1, \\ \mathcal{O}\left[\left(\frac{\partial V_2}{\partial x_1}\right)^*\right] &= 1, \mathcal{O}\left[\left(\frac{\partial V_2}{\partial x_2}\right)^*\right] = 1. \end{aligned} \tag{9.16}$$

Further, using the characteristic values of the velocity components (V_{1C}, V_{2C}) and the characteristic length scales (L_{1C}, L_{2C}) we define the normalized versions of the second derivatives of the velocity components

$$\begin{aligned} \left(\frac{\partial^2 V_1}{\partial x_1 \partial x_1}\right)^* &= \left(\frac{\partial^2 V_1}{\partial x_1 \partial x_1}\right)\frac{L_{1C}^2}{V_{1C}}, \left(\frac{\partial^2 V_1}{\partial x_2 \partial x_2}\right)^* = \left(\frac{\partial^2 V_1}{\partial x_2 \partial x_2}\right)\frac{L_{2C}^2}{V_{1C}}, \\ \left(\frac{\partial^2 V_2}{\partial x_1 \partial x_1}\right)^* &= \left(\frac{\partial^2 V_2}{\partial x_1 \partial x_1}\right)\frac{L_{1C}^2}{V_{2C}}, \left(\frac{\partial^2 V_2}{\partial x_2 \partial x_2}\right)^* = \left(\frac{\partial^2 V_2}{\partial x_2 \partial x_2}\right)\frac{L_{2C}^2}{V_{2C}}, \end{aligned} \tag{9.17}$$

such that

$$\mathcal{O}\left[\left(\frac{\partial^2 V_1}{\partial x_1 \partial x_1}\right)^*\right] = 1, \mathcal{O}\left[\left(\frac{\partial^2 V_1}{\partial x_2 \partial x_2}\right)^*\right] = 1,$$
$$\mathcal{O}\left[\left(\frac{\partial^2 V_2}{\partial x_1 \partial x_1}\right)^*\right] = 1, \mathcal{O}\left[\left(\frac{\partial^2 V_2}{\partial x_2 \partial x_2}\right)^*\right] = 1. \tag{9.18}$$

To perform the order-of-magnitude analysis of (9.3), (9.4), and (9.5), we implement the following three steps separately for each of the three equations:

1. We substitute the dimensional variables in (9.3), (9.4), or (9.5) by their non-dimensional counterparts using (9.13), (9.15), and (9.17).
2. Using the anticipation which is based on the appropriately chosen characteristic values of the velocity components (V_{1C} and V_{2C}) and the length scales (L_{1C} and L_{2C}) that all normalized versions of the variables would have their order-of-magnitude unity over most of the domain of interest, we identify those terms in the respective equations (if any) which have smaller orders of magnitude than the other additive terms of the equation. These identified terms can then be discarded from the equation.
3. We substitute the normalized versions of the variables by their dimensional counterparts using (9.13), (9.15), and (9.17) and arrive at a simplified version of the governing equation.

The order-of-magnitude analysis of the governing equations is performed at an arbitrary location (x_1, x_2) inside the boundary layer region. At such a location, we deem the following choices of the length scales to be appropriate

$$\mathcal{O}[L_{1C}] = \mathcal{O}[x_1],$$
$$\mathcal{O}[L_{2C}] = \mathcal{O}[\delta(x_1)]. \tag{9.19}$$

Further, we deem V_o to be an appropriate choice of V_{1C}. However, at this point, we do not have any cues available to choose V_{2C}. Nonetheless, we do initiate our OM analysis and further explore to see if any further cues emerge to choose V_{2C} appropriately.

9.1.2 The Continuity Equation

Substituting the raw variables by their normalized counterparts in the continuity equation (9.3) results in the following form of the equation:

$$\left(\frac{\partial V_1}{\partial x_1}\right)^* \frac{V_{1C}}{L_{1C}} + \left(\frac{\partial V_2}{\partial x_2}\right)^* \frac{V_{2C}}{L_{2C}} = 0. \tag{9.20}$$

Rearranging various factors, we arrive at

$$\left(\frac{\partial V_1}{\partial x_1}\right)^* + \left(\frac{\partial V_2}{\partial x_2}\right)^* \frac{V_{2C}}{V_{1C}} \frac{L_{1C}}{L_{2C}} = 0. \tag{9.21}$$

The first term on the lhs in (9.21) has its order-of-magnitude unity. Further, since the rhs of (9.21) is zero and there are only two terms on the lhs, the order of magnitude of the second term on the lhs must also be unity.

$$\mathcal{O}\left[\left(\frac{\partial V_2}{\partial x_2}\right)^* \frac{V_{2C}}{V_{1C}} \frac{L_{1C}}{L_{2C}}\right] = 1. \tag{9.22}$$

Since the order of magnitude of the $\left(\frac{\partial V_2}{\partial x_2}\right)^*$ itself is unity, (9.22) leads to the following conclusion:

$$\mathcal{O}\left[\frac{V_{2C}}{V_{1C}} \frac{L_{1C}}{L_{2C}}\right] = 1. \tag{9.23}$$

This, in turn, gives us a cue to choose V_{2C}.

$$V_{2C} = V_{1C} \frac{L_{2C}}{L_{1C}}. \tag{9.24}$$

Thus, even though the order-of-magnitude analysis of the continuity equation does not provide any justification to simplify the equation itself, it has provided a justification based on which we now have a characteristic value of the variable $V_2(x_1, x_2)$ in terms V_{1C}, L_{2C}, and L_{1C} for the flow inside the boundary layer.

9.1.3 The V_1 Equation

We now substitute the raw variables by their normalized counterparts in the governing equation of V_1 (9.4)

$$\left[V_1^* V_{1C}\left(\frac{\partial V_1}{\partial x_1}\right)^* \frac{V_{1C}}{L_{1C}} + V_2^* V_{2C}\left(\frac{\partial V_1}{\partial x_2}\right)^* \frac{V_{1C}}{L_{2C}}\right] =$$
$$g_1 - \frac{1}{\rho}\frac{\partial p}{\partial x_1} + \nu\left(\frac{\partial^2 V_1}{\partial x_1 \partial x_1}\right)^* \frac{V_{1C}}{L_{1C}^2} + \nu\left(\frac{\partial^2 V_1}{\partial x_2 \partial x_2}\right)^* \frac{V_{1C}}{L_{2C}^2}. \tag{9.25}$$

Using (9.24) to substitute V_{2C} in terms of V_{1C}, L_{2C}, and L_{1C} in (9.25) leads to the following form of the equation:

$$\frac{V_{1C}^2}{L_{1C}}\left[V_1^*\left(\frac{\partial V_1}{\partial x_1}\right)^* + V_2^*\left(\frac{\partial V_1}{\partial x_2}\right)^*\right] =$$
$$g_1 - \frac{1}{\rho}\frac{\partial p}{\partial x_1} + \left(\frac{\partial^2 V_1}{\partial x_1 \partial x_1}\right)^* \frac{\nu V_{1C}}{L_{1C}^2} + \left(\frac{\partial^2 V_1}{\partial x_2 \partial x_2}\right)^* \frac{\nu V_{1C}}{L_{2C}^2}. \quad (9.26)$$

The lhs of (9.26) is the stream-wise component of the acceleration of a fluid particle within the boundary layer. The expression within the square parentheses represents the normalized form of this acceleration. Clearly, the OM of this normalized form is unity. Thus, the quantity V_{1C}^2/L_{1C} (the factor outside the square parentheses) can be deemed as the characteristic value of the stream-wise acceleration component. Similarly, on the rhs, the quantities $\nu V_{1C}/L_{1C}^2$ and $\nu V_{1C}/L_{2C}^2$ are the characteristic values of the two viscous force (per unit mass) terms acting on a typical fluid element. Dividing (9.26) throughout by the factor $\frac{V_{1C}^2}{L_{1C}}$ and further re-arranging the terms, we arrive at the following form of the equation:

$$V_1^*\left(\frac{\partial V_1}{\partial x_1}\right)^* + V_2^*\left(\frac{\partial V_1}{\partial x_2}\right)^* = \left(\frac{\partial^2 V_1}{\partial x_1 \partial x_1}\right)^* \frac{\nu}{V_{1C} L_{1C}}$$
$$+ \left(\frac{\partial^2 V_1}{\partial x_2 \partial x_2}\right)^* \frac{\nu}{V_{1C} L_{1C}}\left(\frac{L_{1C}}{L_{2C}}\right)^2 + \frac{g_1 L_{1C}}{V_{1C}^2} - \frac{\partial p}{\partial x_1}\frac{L_{1C}}{\rho V_{1C}^2}. \quad (9.27)$$

The lhs of (9.27) of the equation is the fully normalized form of stream-wise acceleration, with its OM being unity. The rhs of (9.27) represents the sum of the ratios of various forces (per unit mass) to the characteristic value of the stream-wise acceleration of a fluid particle (V_{1C}^2/L_{1C}). If any of these terms on the rhs has its order of magnitude $<< 1$, we can conclude that the particular force is not significant in contributing toward the stream-wise acceleration of the fluid particle. Thus, that term can be neglected in the equation.

The first two terms on the rhs of (9.27) are the ratios of the two viscous force terms to the characteristic value of stream-wise acceleration. Both these terms have a common non-dimensional factor involving ν, V_{1C}, and L_{1C}. We define the inverse of this factor as the *Reynolds number of the governing equation of the stream-wise velocity* component (named after the Irish innovator Osborne Reynolds (1842–1912)). We represent this quantity by the symbol Re_1

$$Re_1 = \frac{V_{1C} L_{1C}}{\nu}. \quad (9.28)$$

Based on this new symbol, the order of magnitude of the first term on the rhs of (9.27) can be expressed as follows:

$$\mathcal{O}\left[\left(\frac{\partial^2 V_1}{\partial x_1 \partial x_1}\right)^* \frac{\nu}{V_{1C} L_{1C}}\right] = \mathcal{O}\left[\frac{\nu}{V_{1C} L_{1C}}\right] = \mathcal{O}\left[\frac{1}{Re_1}\right], \quad (9.29)$$

since $\mathcal{O}\left[\left(\frac{\partial^2 V_1}{\partial x_1 \partial x_1}\right)^*\right] = 1$. The quantity Re_1 is interpreted as the ratio of the characteristic value of the stream-wise acceleration (V_{1C}^2/L_{1C}) of a typical fluid particle to the characteristic value of the viscous force (per unit mass) arising because of the second gradient of V_1 with respect to x_1 $(\nu V_{1C}/L_{1C}^2)$. In line with our overarching aim to focus on flow fields wherein the governing equations can be mathematically simplified, we now restrict ourselves to flow fields which have $Re_1 >> 1$. Accordingly, the first term on the rhs of (9.27) can be neglected.

The factor Re_1 appears in the denominator of the second viscous term, as well, in (9.27). However, unlike the first viscous term, the order of magnitude of the second term is not controlled by Re_1 alone but also by the ratio of two relevant length scales L_{1C} and L_{2C}.

$$\begin{aligned}\mathcal{O}\left[\left(\frac{\partial^2 V_1}{\partial x_2 \partial x_2}\right)^* \frac{\nu}{V_{1C} L_{1C}}\left(\frac{L_{1C}}{L_{2C}}\right)^2\right] &= \mathcal{O}\left[\left(\frac{\nu}{V_{1C} L_{1C}}\right)\left(\frac{L_{1C}}{L_{2C}}\right)^2\right], \\ &= \mathcal{O}\left[\frac{1}{Re_1}\left(\frac{L_{1C}}{L_{2C}}\right)^2\right]. \end{aligned} \tag{9.30}$$

Since the flow inside the boundary layer is indeed affected by viscous forces and further given the fact that when $Re_1 >> 1$, the first viscous term on the rhs of (9.27) is anyway negligible, the second and the only surviving viscous term must be of significance in the evolution equation of V_1. Thus, we conclude that the order of magnitude in (9.30) must itself be unity (same as the order of magnitude of the lhs in (9.27), which represents stream-wise acceleration of a typical fluid particle).

$$\mathcal{O}\left[\frac{1}{Re_1}\left(\frac{L_{1C}}{L_{2C}}\right)^2\right] = 1. \tag{9.31}$$

This in turn allows estimating L_{2C} in terms of other known parameters (L_{1C} and Re_1)

$$\mathcal{O}\left[\frac{L_{2C}}{L_{1C}}\right] = \mathcal{O}\left[\frac{1}{\sqrt{Re_1}}\right]. \tag{9.32}$$

Combining (9.32) with our anticipations in (9.19) we have

$$\mathcal{O}\left[\frac{\delta(x_1)}{x_1}\right] = \mathcal{O}\left[\frac{1}{\sqrt{Re_1}}\right]. \tag{9.33}$$

Equation (9.33) suggests that the boundary layer thickness tends to grow as we move more downstream over the plate. Further, since $Re_1 >> 1$, (9.33) also suggests that the local boundary layer thickness is a small quantity compared to x_1.

The third term on the rhs of (9.27) is a constant involving g_1, V_{1C}, and L_{1C}. We define a non-dimensional quantity called the *Froude number of the governing equation of the stream-wise velocity component,* Fr_1 (named after the English engineer

William Froude, 1810–1879) such that

$$Fr_1 = \sqrt{\frac{V_{1C}^2}{g_1 L_{1C}}}. \tag{9.34}$$

The Froude number squared represents the ratio of the characteristic value of the stream-wise acceleration of a particle to the gravitational force per unit mass in the stream-wise direction. For a flow field with $Fr_1^2 >> 1$, the body force term appearing in the governing equation of V_1 can be neglected.

The fourth term on the rhs of (9.27) represents the ratio of the pressure gradient force per unit mass to the characteristic value of stream-wise acceleration. Since, at this point, we do not have any reasonable estimate of the characteristic value or the order of magnitude of the pressure gradient term itself, we simply retain this term in the V_1 equation. In summary, the order-of-magnitude analysis of the V_1 equation inside the boundary layer flow over a flat plate has led to the following conclusions:

1. The body force term can be neglected if $Fr_1 >> 1$.
2. If $Re_1 >> 1$, then the viscous term arising due to the gradient of the stream-wise velocity component with respect to the stream-wise direction can be neglected, leading to the following simplified form of the V_1 equation

$$V_1 \frac{\partial V_1}{\partial x_1} + V_2 \frac{\partial V_1}{\partial x_2} \approx -\frac{1}{\rho}\frac{\partial p}{\partial x_1} + \nu \frac{\partial^2 V_1}{\partial x_2 \partial x_2} + g_1. \tag{9.35}$$

9.1.4 The V_2 Equation

We substitute the raw variables by their normalized counterparts in the governing equation of V_2 (9.5)

$$\begin{aligned} V_1^* V_{1C} \left(\frac{\partial V_2}{\partial x_1}\right)^* \frac{V_{2C}}{L_{1C}} + V_2^* V_{2C} \left(\frac{\partial V_2}{\partial x_2}\right)^* \frac{V_{2C}}{L_{2C}} &= g_2 - \frac{1}{\rho}\frac{\partial p}{\partial x_2} \\ +\nu \left(\frac{\partial^2 V_2}{\partial x_1 \partial x_1}\right)^* \frac{V_{2C}}{L_{1C}^2} &+ \nu \left(\frac{\partial^2 V_2}{\partial x_2 \partial x_2}\right)^* \frac{V_{2C}}{L_{2C}^2}. \end{aligned} \tag{9.36}$$

Using (9.24) to substitute V_{2C} in terms of V_{1C}, L_{2C}, and L_{1C} in (9.36) leads to the following form of the equation:

$$\begin{aligned} \frac{V_{1C}^2}{L_{1C}}\left(\frac{L_{2C}}{L_{1C}}\right)\left[V_1^*\left(\frac{\partial V_2}{\partial x_1}\right)^* + V_2^*\left(\frac{\partial V_2}{\partial x_2}\right)^*\right] &= \nu \left(\frac{\partial^2 V_2}{\partial x_1 \partial x_1}\right)^* \frac{V_{1C}}{L_{1C}^2}\left(\frac{L_{2C}}{L_{1C}}\right) \\ +\nu \left(\frac{\partial^2 V_2}{\partial x_2 \partial x_2}\right)^* \frac{V_{1C}}{L_{2C}^2}\left(\frac{L_{2C}}{L_{1C}}\right) &+ g_2 - \frac{1}{\rho}\frac{\partial p}{\partial x_2}. \end{aligned} \tag{9.37}$$

The lhs of (9.37) is the wall-normal component of the acceleration of a typical fluid particle within the boundary layer. The expression within the square parentheses represents the normalized form of this acceleration. Clearly, the OM of this normalized form is unity. Thus, the quantity $\left(V_{1C}^2/L_{1C}\right)(L_{2C}/L_{1C})$ (the factor outside the square parentheses) is the characteristic value of the wall-normal acceleration component. Since the characteristic value of stream-wise acceleration is already known to be V_{1C}^2/L_{1C}, the lhs of (9.37) implies that the characteristic value of wall-normal component of the acceleration vector is L_{2C}/L_{1C} times the characteristic value of stream-wise acceleration. The discussion of the previous section has already demonstrated that at $Re_1 >> 1$, L_{2C}/L_{1C} is quite small (9.33). Thus, the net acceleration vector of a typical fluid particle inside the boundary layer is *almost* stream-wise, allowing us to neglect the acceleration term in the V_2 equation and approximate the rest of the equation as a mere force balance in the wall-normal direction.

Further, even within the scope of this approximate force balance, we wish to examine if all the forces (the two viscous terms, the body force term and the pressure gradient term) are individually significant or not. For this analysis, we first divide (9.37) throughout by V_{1C}^2/L_{1C} and arrive at the following form of the V_2 equation:

$$\left[V_1^*\left(\frac{\partial V_2}{\partial x_1}\right)^* + V_2^*\left(\frac{\partial V_2}{\partial x_2}\right)^*\right]\left(\frac{L_{2C}}{L_{1C}}\right) = \left(\frac{\partial^2 V_2}{\partial x_1 \partial x_1}\right)^* \frac{1}{Re_1}\left(\frac{L_{2C}}{L_{1C}}\right)$$
$$+\left(\frac{\partial^2 V_2}{\partial x_2 \partial x_2}\right)^* \frac{1}{Re_1}\left(\frac{L_{1C}}{L_{2C}}\right) + \frac{g_2 L_{1C}}{V_{1C}^2} - \frac{1}{\rho}\frac{\partial p}{\partial x_2}\frac{L_{1C}}{V_{1C}^2}. \tag{9.38}$$

At $Re_1 >> 1$, the order of magnitude of the first term on the rhs is

$$\mathcal{O}\left[\left(\frac{\partial^2 V_2}{\partial x_1 \partial x_1}\right)^* \frac{1}{Re_1}\left(\frac{L_{2C}}{L_{1C}}\right)\right] = \mathcal{O}\left[\frac{1}{Re_1}\left(\frac{L_{2C}}{L_{1C}}\right)\right], \tag{9.39}$$

which is smaller than the lhs of (9.38). The order of magnitude of the second term on the rhs of (9.38) is estimated as

$$\mathcal{O}\left[\left(\frac{\partial^2 V_2}{\partial x_2 \partial x_2}\right)^* \frac{1}{Re_1}\left(\frac{L_{1C}}{L_{2C}}\right)\right] = \mathcal{O}\left[\frac{1}{Re_1}\left(Re_1^{1/2}\right)\right] = \mathcal{O}\left[\frac{1}{Re_1^{1/2}}\right], \tag{9.40}$$

which is the same as the order of magnitude of the lhs of (9.38). However, as discussed before, compared to the order of the magnitude of the stream-wise acceleration, this order of magnitude is much smaller at high Re_1, and thus this viscous term, too, is deemed insignificant. The third term on the rhs of (9.38) is a constant involving g_2, and V_{1C} and L_{1C}. We define a non-dimensional quantity called the *Froude number of the governing equation of the transverse velocity component*, Fr_2.

$$Fr_2 = \sqrt{\frac{V_{1C}^2}{g_2 L_{1C}}}. \tag{9.41}$$

The Froude number squared represents the ratio of the characteristic value of the stream-wise acceleration of a fluid particle to the body force per unit mass in the wall-normal direction. The necessary condition for the body force term to be of any consequence along the wall-normal direction is that the order of magnitude of the third term on the rhs of (9.38) must be greater than the order of magnitude of the lhs of (9.38).

$$\mathcal{O}\left[\frac{g_2 L_{1C}}{V_{1C}^2}\right] > \mathcal{O}\left[\frac{L_{2C}}{L_{1C}}\right], \text{ or } \mathcal{O}\left[\frac{1}{Fr_2^2}\right] > \mathcal{O}\left[\frac{1}{Re_1^{1/2}}\right], \text{ or}$$
$$\mathcal{O}\left[Fr_2^2\right] < \mathcal{O}\left[Re_1^{1/2}\right], \text{ or } \mathcal{O}\left[Fr_2\right] < \mathcal{O}\left[Re_1^{1/4}\right]. \tag{9.42}$$

The fourth term in (9.38) is the ratio of the net pressure force (per unit mass) acting on a particle in the wall-normal direction to the characteristic value of stream-wise acceleration. Since, at this point, we do not have any reasonable estimate of the order-of-magnitude/characteristic value of the pressure gradient term itself, we simply retain this term in the equation.

In summary, the order-of-magnitude analysis of the V_2 equation inside the boundary layer flow over a flat plate has led to the following conclusions:

1. At $Re_1 >> 1$, the wall-normal acceleration is negligible compared to the stream-wise acceleration. Thus, the V_2 equation reduces merely to a balance of forces acting on a fluid particle in the wall-normal direction. Further, all the viscous forces arising due to the gradients of the wall-normal velocity component are inconsequential in this force balance, leading to a mere balance between the pressure force and the body force.

$$-\frac{1}{\rho}\frac{\partial p}{\partial x_2} + g_2 \approx 0. \tag{9.43}$$

2. For the body force term to be of any consequence, the necessary condition is

$$\mathcal{O}\left[Fr_2\right] < \mathcal{O}\left[Re_1^{1/4}\right]. \tag{9.44}$$

We finally assemble the simplified set of governing equations for the flow past a flat plate of a viscous fluid with $Re_1 >> 1$. This set consists of Eqs. (9.3), (9.35), and (9.43).

$$
\begin{aligned}
\frac{\partial V_1}{\partial x_1} + \frac{\partial V_2}{\partial x_2} &= 0, \\
V_1 \frac{\partial V_1}{\partial x_1} + V_2 \frac{\partial V_1}{\partial x_2} &= -\frac{1}{\rho}\frac{\partial p}{\partial x_1} + \nu \frac{\partial^2 V_1}{\partial x_2 \partial x_2} + g_1, \\
-\frac{1}{\rho}\frac{\partial p}{\partial x_2} + g_2 &= 0.
\end{aligned} \tag{9.45}
$$

This is a set of three partial differential equations (PDE) with the dependent variables being V_1, V_2, and p inside the boundary layer. This equation set is also called the *Prandtl's boundary layer equations (PBLE)*. It is named after the German fluid dynamicist Ludwig Prandtl (1875–1953), who was the first to derive this equation set.

To simplify the governing equation further, we further assume that the Froude numbers (Fr_1 and Fr_2) are large enough so that their influences are negligible. Thus, the third equation in the PBLE set (9.45) further simplified to

$$
\frac{1}{\rho}\frac{\partial p}{\partial x_2} \approx 0, \tag{9.46}
$$

which when combined with the earlier conclusion drawn in (9.6) implies that pressure inside the boundary layer must be a function of x_1 alone. In other words, at a fixed x_1 station, pressure must remain unchanged from the wall ($x_2 = 0$) all the way to the edge of the boundary layer ($x_2 = \delta(x_1)$). Since the edge of the boundary layer is actually one of the boundaries of the boundary layer domain over which we would like to solve the PBLE set, the variation of pressure at the edge of the boundary layer must be provided as an input to the differential equation set (9.45). Thus, pressure is no longer a variable of the PBLE set. Accordingly, the pressure gradient term appearing in the V_1 equation can be expressed as

$$
p(x_1, x_2) = p^{edge}(x_1), \tag{9.47}
$$

where $p^{edge}(x_1)$ means the pressure at the edge of the boundary layer at the streamwise location x_1. The PBLE equation set can now be expressed more simply as

$$
\begin{aligned}
\frac{\partial V_1}{\partial x_1} + \frac{\partial V_2}{\partial x_2} &= 0, \\
V_1 \frac{\partial V_1}{\partial x_1} + V_2 \frac{\partial V_1}{\partial x_2} &= -\frac{1}{\rho}\frac{dp^{edge}}{dx_1} + \nu \frac{\partial^2 V_1}{\partial x_2 \partial x_2}.
\end{aligned} \tag{9.48}
$$

This is a set of two PDE in two unknowns V_1 and V_2, where the partial derivative of p^{edge} is the same as the total derivative (p^{edge} is a function of only one independent variable).

The Prandtl boundary layer equation set (9.48) is a simplified version of the Navier–Stokes equation. Nonetheless, it is still a set of partial differential equations.

Even though nowadays, the PDEs of the PBLE set can be easily solved using modern CFD techniques supported by advanced computing hardware, it was not so in the early twentieth century when Prandtl proposed this equation set. In the year 1908, Blassius (a German fluid dynamicist and a student of Ludwig Prandtl) pursued a different approach to solving this equation set. Using the concept of *stream functions* and *self-similar variables*, he first converted the set of PDEs (9.48) into an ordinary differential equation (ODE). Subsequently, he could numerically integrate the ODE to obtain the flow field over the flat plate. In the remaining part of this section, we (i) introduce these two concepts (stream function and self-similar variables), (ii) transform the PBLE equation set into an ODE, (iii) present the numerical solution of the ODE, and (iv) extract some physical insights into the role of viscosity and no-slip condition in shaping the flow behavior inside the boundary layer of a flat plate at high Reynolds number ($Re_1 >> 1$).

9.1.5 Stream Function

In this section, our goal is to define the *stream function* of a 2C, 2D velocity field of a constant density fluid. However, before we present the formal definition of stream functions we consider the control volume shown in Fig. 9.4 and demonstrate that in a constant density fluid with a 2C, 2D velocity field the volumetric flow rate passing through every cross-sectional area of unit depth and bounded by two streamlines (such as cross sections $ABB'A'$ or $DCC'D'$ in Fig. 9.4) is identical. The symbols S_1 and S_2 in this figure are the two streamlines under consideration.

The two cross-sectional areas $ABB'A'$ and $CDD'C'$ are planar. These areas have unit length along the $-\hat{e}_3$ direction. The points A, A', D, and D' lie on the stream surface on which the streamline S_1 lies. The points B, B', C, and C' lie on the stream surface on which the streamline S_2 lies. Using the Reynolds transport theorem, it can be shown that the volumetric flow rates through $ABB'A'$ and $CDD'C'$ are identi-

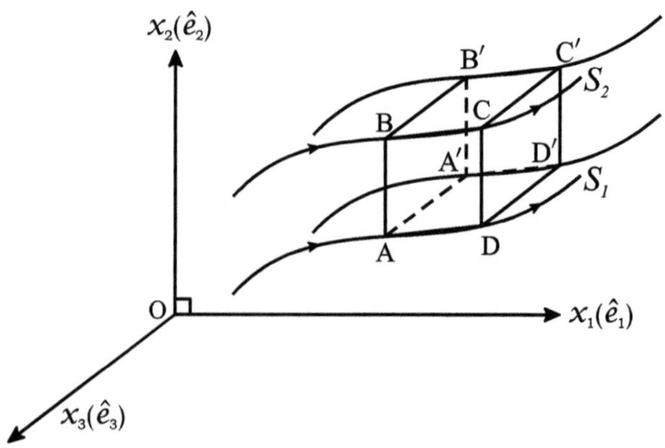

Fig. 9.4 A control volume in a 2C, 2D velocity field

cal. This demonstration requires considering a control volume (fixed to the ground reference frame) comprised of the following six faces: (i) cross section $ABB'A'$ of unit depth, (ii) stream surface $BB'C'C$, (iii) cross section $CDD'C'$ of unit depth, (iv) stream surface $AA'D'D$, (v) area ABCD on the plane $x_1\left(\hat{e}_1\right) - x_2\left(\hat{e}_2\right)$, and (vi) the area $A'B'C'D'$ lying on a plane which is parallel to the $x_1\left(\hat{e}_1\right) - x_2\left(\hat{e}_2\right)$, and is displaced by unit length in the $-\hat{e}_3$ direction. This control volume is shown in Fig. 9.4. We apply (5.12) on this control volume with the entity being mass. Thus, the variable $\underline{\phi}$ appearing (5.12) is a scalar and equals the density of the fluid (ρ). The law of conservation of mass implies that $\underline{\mathcal{R}}_2$ is a scalar and equals zero.

$$
\begin{aligned}
\frac{\partial}{\partial t}\iiint_{\mathcal{V}} \rho d\mathcal{V} + \iint_{\mathcal{S}} \rho\left(\underline{V}\cdot\hat{e}_n\right) dA &= 0, \text{ or}\\
\rho\frac{\partial}{\partial t}\iiint_{\mathcal{V}} d\mathcal{V} + \rho\iint_{\mathcal{S}} \left(\underline{V}\cdot\hat{e}_n\right) dA &= 0, \text{ or}\\
\rho\frac{\partial \mathcal{V}}{\partial t} + \rho\iint_{\mathcal{S}} \left(\underline{V}\cdot\hat{e}_n\right) dA &= 0, \text{ or}\\
\rho\iint_{\mathcal{S}} \left(\underline{V}\cdot\hat{e}_n\right) dA &= 0,
\end{aligned}
\tag{9.49}
$$

where $\mathcal{V}$ represents the volume of the control volume and $\mathcal{S}$ represents the outer surface of it. The symbol $\hat{e}_n$ is the local outward unit normal on the surface of the control volume (CV). The time derivative term has been discarded in (9.49) because the total volume of the considered CV does not change with time. The dot product $\underline{V}\cdot\hat{e}_n$ is identically zero on the two faces $ABCD$ and $A'B'C'D'$, because the velocity field is 2C. Further, the dot product $\underline{V}\cdot\hat{e}_n$ is identically zero on the stream surfaces $BB'C'C$ and $AA'D'D$, because all local velocity vectors on a stream surface must be tangential to the stream surface. Thus, (9.49) simplifies to

$$
\iint_{\mathcal{S}_{\mathcal{ABB'A'}}} \left(\underline{V}\cdot\hat{e}_n\right) dA + \iint_{\mathcal{S}_{\mathcal{CDD'C'}}} \left(\underline{V}\cdot\hat{e}_n\right) dA = 0. \tag{9.50}
$$

Equation (9.50) is alternatively expressed as

$$
-\iint_{\mathcal{S}_{\mathcal{ABB'A'}}} \left(\underline{V}\cdot\hat{e}_n\right) dA = \iint_{\mathcal{S}_{\mathcal{CDD'C'}}} \left(\underline{V}\cdot\hat{e}_n\right) dA, \tag{9.51}
$$

where the surface integrals in (9.50) are to be computed over the surface areas $ABB'A'$ and $CDD'C'$. The left-hand side of (9.51) is the volumetric flow rate passing through the cross section $ABB'A'$ (from left to right), and the rhs of the equation is the volumetric flow rate passing through the cross section $CDD'C'$ (from left to right). Thus, we have proved that in a constant density fluid with a 2C, 2D velocity field, the volumetric flow rate passing through every cross section of unit depth and bounded by a pair of streamlines is identical.

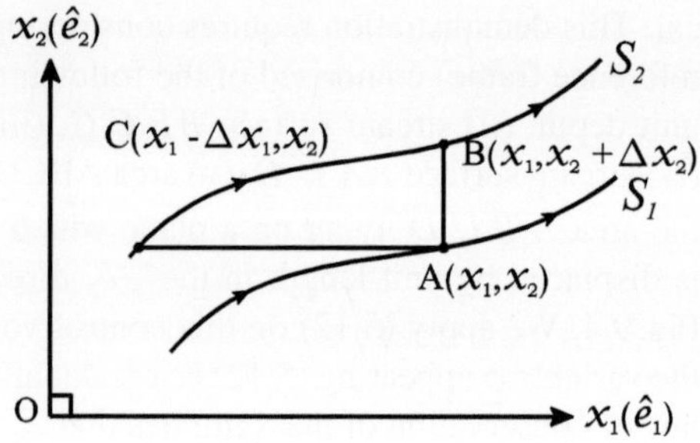

Fig. 9.5 Two infinitesimally spaced streamlines S_1 and S_2

Using the Cartesian coordinate system $Ox_1(\hat{e}_1) - x_2(\hat{e}_2)\, x_3(\hat{e}_3)$ as our working coordinate system, we now define the stream function $\psi(x_1, x_2)$ of a 2C, 2D velocity field of a constant density fluid as

1. $\psi(x_1, x_2) = C_1$ and $\psi(x_1, x_2) = C_2$ are the equations of the two streamline curves (S_1 and S_2) with C_1 and C_2 being two constants.
2. $C_2 - C_1$ must equal the volumetric flow rate per crossing of any arbitrary planar area of unit depth and bounded by the two stream surfaces of which S_1 and S_2 (such as cross sections $ABB'A'$ or $DCC'D'$ in Fig. 9.4).
3. $C_2 > C_1$, if, for an observer placed on the steamline S_1 (Fig. 9.4), the flow direction associated with the streamline S_2 is from left to right.

One obvious utility of defining a stream function is that with the known function, $\psi(x_1, x_2)$, we can readily plot various streamlines in a flow field by setting different numerical values for the constant C in the equation $\psi(x_1, x_2) = C$. Further, we can derive a set of relationships between various partial derivatives of the stream function and the local velocity components of a 2C, 2D velocity field. To derive these relationships, we refer to Fig. 9.5. In this figure, we show two streamlines S_1 (equation of the curve $\psi(x_1, x_2) = C_1$) and S_2 (equation of the curve $\psi(x_1, x_2) = C_2$), which are infinitesimally close to each other with the direction of the velocity field indicated by arrows on the two streamlines. We consider three locations A, B, and C. The location A lies on streamlines S_1, whereas locations B and C lie on the streamline S_2. The locations have been identified such that the line segment AB is parallel to $\hat{e}_2$ and the line segment CA is parallel to $\hat{e}_1$. The Cartesian coordinates of location A are (x_1, x_2), whereas those of B and C are $(x_1, x_2 + \Delta x_2)$ and $(x_1 - \Delta x_1, x_2)$, respectively. The symbols $\Delta x_1 (> 0)$ and $\Delta x_2 (> 0)$ represent infinitesimal increments in x_1 and x_2. For small, arbitrary and independent changes in Δx_1 and Δx_2, the corresponding change in the function ψ is expressed as

$$\Delta\psi \approx \frac{\partial \psi}{\partial x_1}\Delta x_1 + \frac{\partial \psi}{\partial x_2}\Delta x_2. \tag{9.52}$$

If we now consider the pair of locations A and B as shown in Fig. 9.5, (9.52) simplifies to the following approximate form:

$$\Delta\psi = \psi(x_1, x_2 + \Delta x_2) - \psi(x_1, x_2) \approx \frac{\partial \psi}{\partial x_2}\Delta x_2. \tag{9.53}$$

Since the points B and A lie on the streamlines S_2 and S_1, respectively,

$$\psi(x_1, x_2 + \Delta x_2) = C_2, \tag{9.54}$$

$$\psi(x_1, x_2) = C_1. \tag{9.55}$$

Subtracting (9.55) from (9.54) we get

$$\psi(x_1, x_2 + \Delta x_2) - \psi(x_1, x_2) = C_2 - C_1. \tag{9.56}$$

Equating the rhs of (9.53) to the rhs of (9.56) leads to

$$\frac{\partial \psi}{\partial x_2}\Delta x_2 \approx C_2 - C_1. \tag{9.57}$$

By definition, the rhs of (9.57) equals the volumetric flow rate per unit depth through the cross-sectional area corresponding to the edge AB. This flow rate can be expressed in terms of local velocity component, V_1, as

$$C_2 - C_1 \approx V_1(x_1, x_2)\Delta x_2. \tag{9.58}$$

Now equating the rhs of (9.58) to the lhs of (9.57) in the limit $\Delta x_2 \rightarrow 0$ leads to the following exact relationship:

$$V_1(x_1, x_2) = \frac{\partial \psi}{\partial x_2}. \tag{9.59}$$

To find a similar relationship between V_2 and a local partial derivative of ψ, we consider the pair of locations A and C (Fig. 9.5), which have coordinates (x_1,x_2) and ($x_1 - \Delta x_1$,x_2), respectively. Equation (9.52) simplifies to

$$\psi(x_1 - \Delta x_1, x_2) - \psi(x_1, x_2) \approx \frac{\partial \psi}{\partial x_1}(-\Delta x_1). \tag{9.60}$$

Since the points C and A lie on the streamlines S_2 and S_1, respectively,

$$\psi(x_1 - \Delta x_1, x_2) - \psi(x_1, x_2) = C_2 - C_1. \tag{9.61}$$

Equating the right-hand sides of (9.60) and (9.61) leads to

$$C_2 - C_1 \approx \frac{\partial \psi}{\partial x_1}(-\Delta x_1). \tag{9.62}$$

By definition of stream function, the lhs of (9.62) equals the volumetric flow rate per unit depth through the cross-sectional area corresponding to the edge CA. This flow rate can be expressed in terms of the local velocity component, V_2, as

$$C_2 - C_1 \approx V_2(x_1, x_2)\Delta x_1. \tag{9.63}$$

Equating the rhs of (9.63) to the rhs of (9.62) in the limit $\Delta x_1 \to 0$ leads to the following exact relationship:

$$V_2(x_1, x_2) = -\frac{\partial \psi}{\partial x_1}. \tag{9.64}$$

Even though, in this section, we have presented the definition of stream functions with the immediate purpose to implement the Blassius' transformation of the PBLE set, this definition itself and the associated relationships with the local velocity components (Eqs. 9.59 and 9.64) are applicable to all 2C, 2D velocity fields of a constant density real, as well as, ideal fluid.

9.1.6 Blassius Solution

To transform Prandtl's boundary layer equation set (9.48) to an ordinary differential equation, Blassius designated the stream function $\psi(x_1, x_2)$ as the single dependent variable to substitute for V_1 and V_2 (Eqs. 9.59 and 9.64). With this substitution, the continuity equation in the set (9.45) is unconditionally satisfied.

$$\frac{\partial V_1}{\partial x_1} + \frac{\partial V_2}{\partial x_2} = \frac{\partial}{\partial x_1}\left(\frac{\partial \psi}{\partial x_2}\right) + \frac{\partial}{\partial x_2}\left(-\frac{\partial \psi}{\partial x_1}\right) = \frac{\partial^2 \psi}{\partial x_1 \partial x_2} - \frac{\partial^2 \psi}{\partial x_2 \partial x_1} = 0. \tag{9.65}$$

Thus, now the PBLE equation set (9.48) reduces to one PDE

$$V_1\frac{\partial V_1}{\partial x_1} + V_2\frac{\partial V_1}{\partial x_2} = -\frac{1}{\rho}\frac{dp^{edge}}{dx_1} + \nu\frac{\partial^2 V_1}{\partial x_2 \partial x_2}. \tag{9.66}$$

Next, to comprehend what may control the pressure variation at the edge of the boundary layer, we refer to Fig. 9.3. As discussed earlier, this flow domain can be perceived as two regions separated by the curve that describes the edge of the boundary layer, $\delta(x_1)$. The region bounded by the solid plate and the boundary layer edge is indeed the boundary layer region, wherein, by definition, the viscous forces are significant enough to influence the acceleration of a typical fluid particle therein. This demarcation implies that the region outside the boundary layer has a negligible influence of viscous forces on the acceleration of the fluid particles that are outside the boundary layer. Thus, in this outer region, the flow must be governed by the steady Euler equations with the velocity field being 2C and 2D (8.1). Further, since the velocity field in the far-upstream region is uniform and lacks any vorticity therein, the velocity field outside the boundary layer may be approximated to be potential (explanation available in Sect. 8.4). This, in turn, implies that the generalized Bernoulli Equation (8.40) can be applied between two locations in the region outside the boundary layer. Choosing location A to be on the $x_1\hat{e}_1$ axis in the far-upstream region where both the velocity ($V_o\hat{e}_1$) and pressure (p_o) are known and choosing B to be located at the edge of the boundary layer at station x_1, we have the following relationship:

$$p^{edge}(x_1) = p_o + \frac{\rho V_0^2}{2} - \frac{\rho}{2}\left(V^2\right)_{(x_1,\delta(x_1))}, \tag{9.67}$$

where $\left(V^2\right)_{(x_1,\delta(x_1))}$ is the square of the magnitude of local velocity at the edge of the boundary layer at station x_1 (Fig. 9.3). The influence of body forces has already been assumed to be negligible in (9.67).

At the edge of the boundary layer, we have

$$\left(V^2\right)_{(x_1,\delta(x_1))} = \left(V_1^2\right)_{(x_1,\delta(x_1))} + \left(V_2^2\right)_{(x_1,\delta(x_1))}. \tag{9.68}$$

However, based on the OM analysis, we are aware that at $Re_1 >> 1$ (Eqs. 9.24 and 9.32), we have

$$\mathcal{O}\left(\frac{V_2}{V_1}\right) = \mathcal{O}\left(\frac{V_{2C}}{V_{1C}}\right) = \mathcal{O}\left(\frac{L_{2C}}{L_{1C}}\right) = \mathcal{O}\left(\frac{1}{\sqrt{Re_1}}\right) << 1. \tag{9.69}$$

Thus, we approximate (9.68) to

$$\left(V^2\right)_{(x_1,\delta(x_1))} \approx \left(V_1^2\right)_{(x_1,\delta(x_1))}. \tag{9.70}$$

Further, by using the definition of $\delta(x_1)$ (9.1) to express V_1 at the edge of the boundary layer in terms of V_o, Eq. (9.70) is expressed as

$$\left(V^2\right)_{(x_1,\delta(x_1))} \approx (0.99V_o)^2. \tag{9.71}$$

Thus, (9.67) is expressed as

$$p^{edge}(x_1) \approx p_o + \frac{\rho 0.02(V_o)^2}{2}, \tag{9.72}$$

where the rhs is no more a function of x_1. Thus,

$$\frac{dp^{edge}}{dx_1} \approx 0. \tag{9.73}$$

Blassius introduced this simplification in (9.66) to simplify the V_1 equation as

$$V_1 \frac{\partial V_1}{\partial x_1} + V_2 \frac{\partial V_1}{\partial x_2} = \nu \frac{\partial^2 V_1}{\partial x_2 \partial x_2}. \tag{9.74}$$

To further simplify (9.74), Blassius invoked the concept of *self-similarity*. A dependent variable having s number of independent variables is said to exhibit self-similarity, if the dependence on the independent variables can be completely described by m number of combinations of s independent variables, where $m < s$. Blassius recognized that for the special case of the boundary layer over a flat plate for which $\frac{dp^{edge}}{dx_1} = 0$, a self-similar dependent variable $f(\eta)$ can be defined as

$$f(\eta) = \frac{\psi}{\sqrt{\nu x_1 V_o}}, \tag{9.75}$$

where

$$\eta = x_2 \sqrt{\frac{V_o}{\nu x_1}}, \tag{9.76}$$

and ψ is the stream function of the flow field. In the original form of the PBLE set, we have two independent variables: x_1 and x_2. However, now a new variable η is being proposed which is a combination of x_1 and x_2. The utility of recognizing such a new variable is justifiable because it can be demonstrated that a subsequent attempt to substitute the original variables (V_1, V_2, x_1, and x_2) in terms of η, $f(\eta)$ and derivatives of $f(\eta)$ with respect to η does indeed lead to the transformation of the partial differential equation (9.74) into an ordinary differential equation with $f(\eta)$ being the only dependent variable and η being the only independent variable. The algebraic steps involved in this transformation are illustrated in the following equations:

$$\begin{aligned} V_1 &= \frac{\partial \psi}{\partial x_2} = \frac{\partial}{\partial x_2}\left[f(\eta)\sqrt{\nu x_1 V_o}\right] = \sqrt{\nu x_1 V_o}\frac{\partial f}{\partial x_2} = \sqrt{\nu x_1 V_o}\frac{df}{d\eta}\frac{\partial \eta}{\partial x_2}, \\ &= \sqrt{\nu x_1 V_o}\frac{df}{d\eta}\left[\sqrt{\frac{V_o}{\nu x_1}}\right] = V_o\frac{df}{d\eta} = V_o f^{'}(\eta), \end{aligned} \tag{9.77}$$

where $f^{'}(\eta) = \dfrac{df}{d\eta}$.

$$\begin{aligned} V_2 &= -\frac{\partial \psi}{\partial x_1} = -\frac{\partial}{\partial x_1}\left[f\sqrt{\nu x_1 V_o}\right] = -\frac{\partial f}{\partial x_1}\sqrt{\nu x_1 V_o} - \frac{f\sqrt{\nu V_o}}{2\sqrt{x_1}}, \\ &= -\frac{df}{d\eta}\frac{\partial \eta}{\partial x_1}\sqrt{\nu x_1 V_o} - \frac{f}{2}\sqrt{\frac{\nu V_o}{x_1}}, \end{aligned}$$

$$
\begin{aligned}
&= \frac{df}{d\eta}\left(x_2\sqrt{\frac{V_o}{\nu}}\frac{1}{2x_1\sqrt{x_1}}\right)\sqrt{\nu x_1 V_o} - \frac{f}{2}\sqrt{\frac{\nu V_o}{x_1}}, \\
&= \frac{f^{'}}{2}\left(x_2\sqrt{\frac{V_o}{\nu x_1}}\frac{1}{x_1}\right)\sqrt{\nu x_1 V_o} - \frac{f}{2}\sqrt{\frac{\nu V_o}{x_1}}, \\
&= \frac{f^{'}}{2}\eta\sqrt{\frac{\nu V_o}{x_1}} - \frac{f}{2}\sqrt{\frac{\nu V_o}{x_1}}, \\
&= \frac{1}{2}\sqrt{\frac{\nu V_o}{x_1}}\left[\eta f^{'} - f\right].
\end{aligned} \tag{9.78}
$$

We use the expressions derived in (9.77) and (9.78) to find the derivatives of V_1.

$$
\begin{aligned}
\frac{\partial V_1}{\partial x_1} &= \frac{\partial}{\partial x_1}\left[V_o f^{'}(\eta)\right] = V_o\frac{\partial f^{'}}{\partial x_1} = V_o\frac{df^{'}}{d\eta}\frac{\partial \eta}{\partial x_1}, \\
&= V_o\frac{df^{'}}{d\eta}x_2\sqrt{\frac{V_o}{\nu}}\left(-\frac{1}{2x_1\sqrt{x_1}}\right), \\
&= V_o f^{''}\left(x_2\sqrt{\frac{V_o}{\nu x_1}}\right)\left(\frac{-1}{2x_1}\right) = \frac{-V_o\eta f^{''}}{2x_1},
\end{aligned} \tag{9.79}
$$

where $f^{''} = \frac{d^2 f}{d\eta^2}$. Similarly,

$$
\begin{aligned}
\frac{\partial V_1}{\partial x_2} &= \frac{\partial}{\partial x_2}\left[V_o f^{'}\right] = V_o\frac{\partial f^{'}}{\partial x_2} = V_o\frac{df^{'}}{d\eta}\frac{\partial \eta}{\partial x_2} = V_o f''\frac{\partial}{\partial x_2}\left[x_2\sqrt{\frac{V_o}{\nu x_1}}\right], \\
&= V_o f^{''}\sqrt{\frac{V_o}{\nu x_1}}.
\end{aligned} \tag{9.80}
$$

Further, we take a higher derivative of (9.80) with respect to x_2

$$
\begin{aligned}
\frac{\partial^2 V_1}{\partial x_2 \partial x_2} &= \frac{\partial}{\partial x_2}\left[V_o f^{''}\sqrt{\frac{V_o}{\nu x_1}}\right] = V_o\sqrt{\frac{V_o}{\nu x_1}}\frac{df^{''}}{d\eta}\frac{\partial \eta}{\partial x_2}, \\
&= V_o\sqrt{\frac{V_o}{\nu x_1}}f^{'''}\sqrt{\frac{V_o}{\nu x_1}} = \frac{V_o^2 f^{'''}}{\nu x_1},
\end{aligned} \tag{9.81}
$$

where $f^{'''} = \frac{d^3 f}{d\eta^3}$, and all partial derivatives of f with respect to η have been expressed as

$$
\frac{\partial f(\eta)}{\partial \eta} = \frac{df}{d\eta}, \; \frac{\partial^2 f(\eta)}{\partial \eta^2} = \frac{d^2 f}{d\eta^2}, \; \frac{\partial^3 f(\eta)}{\partial \eta^3} = \frac{d^3 f}{d\eta^3} \tag{9.82}
$$

since $f(\eta)$ is a self-similar function of η alone. Using (9.77)–(9.81) in (9.74) transforms the PDE into an ordinary algebraic equation (ODE)

$$f''' + \frac{f'' f}{2} = 0. \tag{9.83}$$

Equation (9.83) is a third-order ODE in one dependent variable (f) and one independent variable (η). This equation can be numerically integrated using the following three boundary conditions:

1. No-slip condition at the wall:

$$V_1(x_1 > 0, x_2 = 0) = 0 \Rightarrow f'(\eta = 0) = 0.$$

2. No-penetration condition at the wall:

$$\begin{aligned} V_2(x_1 > 0, x_2 = 0) = 0 &\Rightarrow \eta f'(\eta = 0) - f(\eta = 0) = 0, \\ &\Rightarrow f(\eta = 0) = 0. \end{aligned}$$

3. At $x_1 > 0$:

$$\begin{aligned} &\text{as } x_2 \to \infty, V_1(x_1, x_2) \to V_o, \\ &\Rightarrow \text{as } \eta \to \infty, V_o f' \to V_o, \\ &\Rightarrow f' \to 1. \end{aligned}$$

Table 9.1 presents the approximate numerical solution of (9.83). This is called the Blassius solution of the Prandtl boundary layer equation for the flow over a flat plate. In Table 9.1, the independent column is that of η in the range $0 \le \eta \le 6.00$. We have presented the solution of (9.83) in terms of f, f', and f''. The other columns include the corresponding values of f, f', and f''. We observe that f, f' increase monotonically as η increases. In contrast, f'' has its maximum value at $\eta = 0$, and it monotonically decreases as η increases. The quantity f' is the ratio of local stream-wise velocity to the far-upstream velocity (V_1/Vo). We observe that f' is approximately 0.99 at $\eta \approx 5.0$. A further increase in η tends to bring f' closer to unity. Using the definition of the boundary layer thickness (9.1) and the definition of η (9.76), we conclude that at an x_1 station, the boundary layer thickness equals x_2 such that

Table 9.1 Approximate numerical solution of the ordinary differential equation (9.83) for the flow over a flat plate

η	f	f'	f''	η	f	f'	f''
0.00	0.00000	–0.00000	0.33206	3.00	1.39681	0.84604	0.16136
0.20	0.00664	0.06641	0.33198	3.20	1.56909	0.87608	0.13913
0.40	0.02656	0.13276	0.33147	3.40	1.74695	0.90176	0.11788
0.60	0.05973	0.19894	0.33008	3.60	1.92953	0.92333	0.09809
0.80	0.10611	0.26471	0.32739	3.80	2.11603	0.94112	0.08013
1.00	0.16557	0.32978	0.32301	4.00	2.30575	0.95552	0.06423
1.20	0.23795	0.39378	0.31659	4.20	2.49804	0.96696	0.05052
1.40	0.32298	0.45626	0.30787	4.40	2.69236	0.97587	0.03897
1.60	0.42032	0.51676	0.29666	4.60	2.88825	0.98268	0.02948
1.80	0.52952	0.57476	0.28293	4.80	3.08532	0.98779	0.02187
2.00	0.65002	0.62977	0.26675	5.00	3.28327	0.99154	0.01591
2.20	0.78119	0.68131	0.24835	5.20	3.48187	0.99425	0.01134
2.40	0.92229	0.72898	0.22809	5.40	3.68092	0.99616	0.00793
2.60	1.07251	0.77246	0.20645	5.60	3.88029	0.99748	0.00543
2.80	1.23098	0.81151	0.18401	5.80	4.07988	0.99838	0.00365
				6.00	4.27962	0.99897	0.00240

$$\eta \approx 5.00,$$
$$\Rightarrow \delta(x_1)\sqrt{\frac{V_o}{\nu x_1}} \approx 5.00 \Rightarrow \delta(x_1) \approx 5.00\sqrt{\frac{\nu x_1}{V_o}},$$
$$\Rightarrow \frac{\delta(x_1)}{x_1} \approx \frac{5.00}{\sqrt{Re_1}}. \tag{9.84}$$

The Blassius solution clearly shows that $\delta(x_1) << x_1$ at $Re_1 >> 1$, the boundary layer is indeed a thin zone compared to the magnitude of x_1. Further, (9.84) shows that as x_1 increases, the boundary layer thickness monotonically increases.

Using (9.77) and (9.78) in conjunction with the numerical solution of the ODE (9.83) available in Table 9.1, we can compute the velocity components V_1 and V_2 at any location within the boundary layer where the condition $Re_1 >> 1$ is satisfied. For illustration, we choose typical values of the constants ν and V_o: $\nu = 8.947 \times 10^{-7}$ m^2s^{-1} (water at 300 K) and $V_o = 1$ m/s. In Fig. 9.6, we plot the variation of V_1/V_o versus x_2 over the range $x_2 = 0$ and $x_2 = \delta(x_1)$ at three representative x_1 stations: $x_1 = 0.1$ m, $x_1 = 0.2$ m, and $x_1 = 0.3$ m. The corresponding Re_1 at these x_1 stations are $1.1 \times 10^5, 2.2 \times 10^5$ and 3.3×10^5, which are all $>> 1$. All curves satisfy the no-slip condition at $x_2 = 0$ $(V_1 = 0)$. Further, at $x_2 = \delta(x_1)$, $V_1 \approx 0.99$. Each velocity profile shows a monotonic, albeit non-linear, variation with x_2.

We plot the variation of V_2/V_o in the wall-normal direction at the same three x_1 stations. We observe that as x_2 increases, V_2 first increases but then tends to reach an

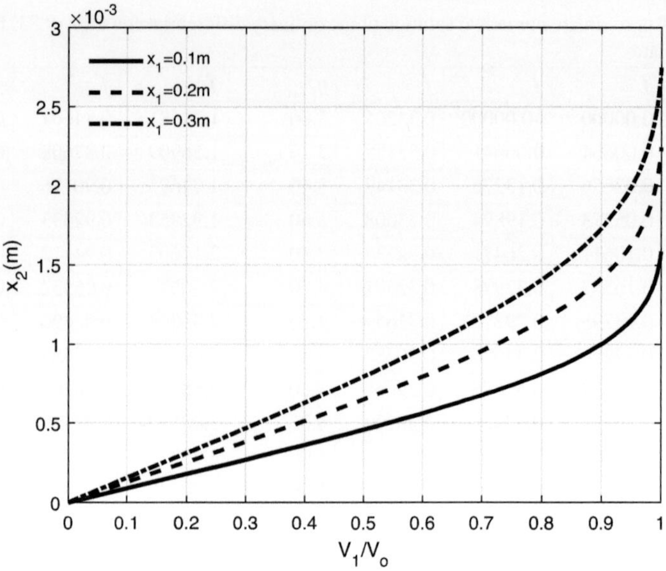

Fig. 9.6 Blassius solution: stream-wise velocity profiles at various x_1 stations

asymptotic state as one approaches the edge of the boundary layer at the respective x_1 station. Further, we observe that even though V_2 is non-zero (everywhere except at the wall), its magnitude is much smaller than the corresponding magnitude of V_1 (Fig. 9.6). It can be verified that this disparity in the magnitudes of the two velocity components is indeed in accordance with (9.24). Since V_2 within the boundary layer is not exactly zero, the direction of the local velocity vector must be inclined to the direction of $\hat{e}_1$. If θ represents the inclination of the local velocity vector relative to $\hat{e}_1$, then the following relationship can be derived (Fig. 9.7).

$$\theta = tan^{-1}\left(\frac{V_2}{V_1}\right) = tan^{-1}\left[\frac{1}{2}\sqrt{\frac{1}{Re_1}}\left(\eta - \frac{f}{f'}\right)\right], \tag{9.85}$$

where expressions from (9.77) and (9.78) have been used to substitute V_1 and V_2 in terms of f and its derivatives. Specifically, at the edge of the boundary layer, $\left(\eta - \frac{f}{f'}\right) \approx 1.7$ (Table 9.1), and thus

$$\theta = tan^{-1}\left(\frac{V_2}{V_1}\right) = tan^{-1}\left[0.85\sqrt{\frac{1}{Re_1}}\right]. \tag{9.86}$$

On the other hand, the boundary layer thickness continues to grow with x_1. The local slope of the boundary layer edge can be determined by taking the derivative of (9.84)

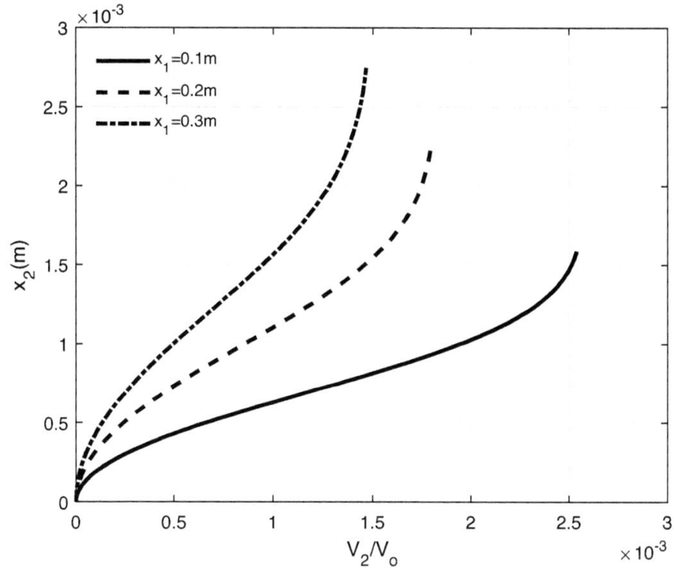

Fig. 9.7 Blassius solution: wall-normal velocity profiles at various x_1 stations

with respect to x_1

$$\beta = tan^{-1}\left[\frac{d\delta(x_1)}{dx_1}\right] = tan^{-1}\left[2.5\sqrt{\frac{1}{Re_1}}\right], \tag{9.87}$$

where β is the angle that the local tangent to the edge of the boundary layer makes with the direction of $\hat{e}_1$ (Fig. 9.8).

A comparison of (9.86) and (9.87) clearly shows that $\beta > \theta$. This, in turn, implies that at all locations on the edge of the boundary layer, there is a net influx of mass from the outer potential region into the boundary layer to the inside of the boundary layer (and not vice versa).

Using the expressions of V_1, V_2 and their spatial derivatives in terms of f and their derivatives in (9.77)–(9.81) and the numerical solution available in Table 9.1, it can be verified that the following relationships are valid inside the boundary layer:

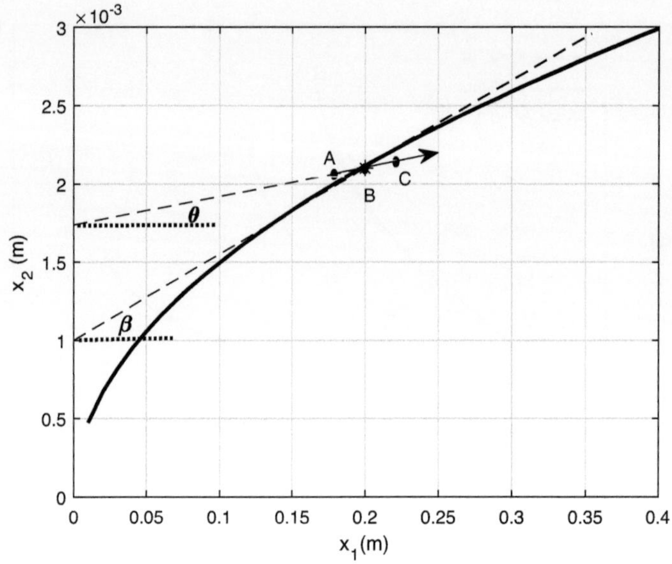

Fig. 9.8 Orientations of the local velocity vector (arrow) and the local tangent to $\delta(x_1)$ at $x_1 = 0.2$ m. The solid curve represents the boundary layer edge (using the representative values: $\nu = 8.947 \times 10^{-7}$ m^{2}s^{-1} and $V_o = 1$ m/s). The dotted line segments show the orientation of $\hat{e}_1$

$$
\begin{aligned}
&\mathcal{O}[V_1] = \mathcal{O}[V_o], \mathcal{O}[V_2] = \mathcal{O}\left[\frac{V_o\delta(x_1)}{x_1}\right] = \mathcal{O}\left[\frac{V_o}{\sqrt{Re_1}}\right],\\
&\mathcal{O}\left[\frac{\partial V_1}{\partial x_1}\right] = \mathcal{O}\left[\frac{V_o}{x_1}\right], \mathcal{O}\left[\frac{\partial V_1}{\partial x_2}\right] = \mathcal{O}\left[\frac{V_o}{\delta(x_1)}\right] = \mathcal{O}\left[\frac{V_o\sqrt{Re_1}}{x_1}\right],\\
&\mathcal{O}\left[\frac{\partial V_2}{\partial x_1}\right] = \mathcal{O}\left[\frac{V_o\delta(x_1)}{x_1^2}\right] = \mathcal{O}\left[\frac{V_o}{x_1\sqrt{Re_1}}\right],\\
&\mathcal{O}\left[\frac{\partial V_2}{\partial x_2}\right] = \mathcal{O}\left[\frac{V_o\delta(x_1)}{x_1\delta(x_1)}\right] = \mathcal{O}\left[\frac{V_o}{x_1}\right].
\end{aligned} \tag{9.88}
$$

The Blassius solution can be used to find the stress in the fluid at the plate interface $(x_1, x_2 = 0)$. The fluid being Newtonian, the full expression of the stress tensor (with reference to the working Cartesian coordinate system shown in Fig. 9.1) is

$$
\begin{aligned}
(\underline{\sigma})_{x_1,x_2=0} = (\tau_{21})_{x_1,x_2=0}\,\hat{e}_2\hat{e}_1 + (-p + \tau_{22})_{x_1,x_2=0}\,\hat{e}_2\hat{e}_2\\
+ (\tau_{23})_{x_1,x_2=0}\,\hat{e}_2\hat{e}_3.
\end{aligned} \tag{9.89}
$$

On the other hand, the pressure stress is constant on the plate at all x_1 stations, and it approximately equals $p^{edge}(x_1)$ (9.47) and (9.72). The stress component $\tau_{23} = \mu\left(\frac{\partial V_3}{\partial x_2} + \frac{\partial V_3}{\partial x_2}\right)$ must be zero, because the velocity field is 2C and 2D. Further, at $Re_1 >> 1$ (9.88)

$$\mathcal{O}\left[\frac{\partial V_1}{\partial x_2}\right] = \mathcal{O}\left[\frac{V_o\sqrt{Re_1}}{x_1}\right] \text{ and } \mathcal{O}\left[\frac{\partial V_2}{\partial x_1}\right] = \mathcal{O}\left[\frac{V_o}{x_1\sqrt{Re_1}}\right]. \tag{9.90}$$

Clearly, when $Re_1 >> 1$

$$\frac{\partial V_1}{\partial x_2} >> \frac{\partial V_2}{\partial x_1}. \tag{9.91}$$

Accordingly, the shear stress component τ_{21} can be approximated as

$$\tau_{21} = \mu\left(\frac{\partial V_1}{\partial x_2} + \frac{\partial V_2}{\partial x_1}\right) \approx \mu\frac{\partial V_1}{\partial x_2}. \tag{9.92}$$

Further, (9.88) shows that

$$\mathcal{O}\left[\frac{\partial V_2}{\partial x_2}\right] = \mathcal{O}\left[\frac{V_o}{x_1}\right], \tag{9.93}$$

which is much smaller than the order of magnitude of $\frac{\partial V_1}{\partial x_2}$ (9.90). Thus,

$$\tau_{22} = 2\mu\frac{\partial V_2}{\partial x_2} << \mu\frac{\partial V_1}{\partial x_2} \approx \tau_{21}. \tag{9.94}$$

Accordingly, the local stress tensor (9.89) is approximated as

$$(\underline{\sigma})_{x_1,x_2=0} \approx \mu\left(\frac{\partial V_1}{\partial x_2}\right)_{x_1,x_2=0} \hat{e}_2\hat{e}_1 - p^{edge}\hat{e}_2\hat{e}_2. \tag{9.95}$$

Using the expression in (9.80) and the numerical solution available in Table 9.1, the shear stress at the station x_1 on the plate $(x_1, x_2 = 0)$ is

$$\left(\mu\frac{\partial V_1}{\partial x_2}\right)_{x_1,x_2=0} = \mu V_o\sqrt{\frac{V_o}{\nu x_1}}f''(\eta = 0) = 0.33\mu V_o\sqrt{\frac{V_o}{\nu x_1}}. \tag{9.96}$$

The expression shows that the shear stress in the fluid at the wall interface decreases as one moves to a more downstream station on the plate. Since $\frac{\partial V_1}{\partial x_2} > 0$ at the wall at all x_1 stations, the integrated shear force on the entire plate area wetted by the fluid is a force along $\hat{e}_1$. On the other hand, the net force along the $-\hat{e}_2$ direction of the plate is simply the product of the plate area and p^{edge}. There must be a balancing force being exerted by an external agency on the plate to keep it fixed with respect to the ground frame as the fluid flows over it.

Using the Blassius solution, we can also examine the vorticity field existing inside a flat plate boundary layer. For a 2C, 2D flow field, the only non-trivial component of the vorticity vector is ω_3 (4.38)

$$\omega_3 = \left(\frac{\partial V_2}{\partial x_1} - \frac{\partial V_1}{\partial x_2}\right). \tag{9.97}$$

Using the approximation (9.91) and then subsequently using (9.80), (9.97) simplifies to

$$\omega_3 \approx -\frac{\partial V_1}{\partial x_2} = -V_o\sqrt{\frac{V_o}{\nu x_1}}f''. \tag{9.98}$$

The known variation of f'' versus η from Table 9.1 suggests that

1. $\omega_3 < 0$ everywhere inside the boundary layer. This, in turn, implies that all small fluid elements inside the boundary have clockwise averaged angular velocity.
2. At a fixed x_1 station, the magnitude of vorticity is maximum at the wall ($x_2 = 0$). The no-slip condition exiting at the wall surface inherently gives rise to this non-zero vorticity. Further, the magnitude of ω_3 monotonically decreases and tends to vanish as one moves toward the edge of the boundary layer (f'' decreases monotonically, as $\eta \to 5.0$ in Table 9.1).

Referring to Fig. 9.8, let us focus on the shown streamline segment ABC. The location A lies in the outer potential flow region. Location B is on the edge of the boundary layer. Location C is inside the boundary layer region. A small fluid element which follows this streamline has no vorticity when it is at a far-upstream location. It continues its journey outside the boundary layer, which can be approximated as a potential flow region. This region still remains free of vorticity. However, once it crosses the edge of the boundary layer and enters the boundary layer region, its vorticity starts evolving. The cause of this vorticity evolution is the presence of viscous effects inside the boundary layer. Earlier in Sect. 8.3, we derived the vorticity Equation (8.29) by taking the curl of the Euler equation (assuming the fluid to be an ideal fluid). This derivation process can be extended to the flow of a real fluid as well by adding the curl of the viscous force term appearing in the exact governing equation of $\underline{V}$ (5.110). The resulting exact transport equation of the vorticity vector in a general 3C, 3D flow field is

$$\frac{D\underline{\omega}}{Dt} = \frac{1}{\rho}\left(\underline{\omega}\cdot\underline{\nabla}\right)\underline{V} + \nu\nabla^2\underline{\omega}. \tag{9.99}$$

For a 2C, 2D velocity field, the only non-trivial scalar equation that can be extracted from the vector Equation (9.99) is that of ω_3. Further, it can be verified that in that equation the vortex-stretching process $\left[\left(\underline{\omega}\cdot\underline{V}\right)\underline{V}\right]$ is zero leading to the following simple equation:

$$\frac{D\omega_3}{Dt} = \nu\nabla^2\omega_3 = \nu\left(\frac{\partial^2\omega_3}{\partial x_1^2} + \frac{\partial^2\omega_3}{\partial x_2^2}\right). \tag{9.100}$$

Now using the approximations that $\omega_3 \approx -\frac{\partial V_1}{\partial x_2}$, which is valid inside a flat plate boundary layer at $Re_1 >> 1$, we arrive at

$$\frac{D\omega_3}{Dt} = -\nu\left[\frac{\partial^3 V_1}{\partial x_1^2\partial x_2} + \frac{\partial^3 V_1}{\partial x_2^3}\right]. \tag{9.101}$$

The orders of magnitude of the third derivatives appearing in (9.101) are estimated using the characteristic velocity values and the characteristic lengths scales as

$$\mathcal{O}\left[\frac{\partial^3 V_1}{\partial^2 x_1 \partial x_2}\right] = \mathcal{O}\left[\frac{V_o}{x_1^2 \delta(x_1)}\right] = \mathcal{O}\left[\frac{V_o Re^{1/2}}{x_1^3}\right],$$

$$\mathcal{O}\left[\frac{\partial^3 V_1}{\partial^3 x_2}\right] = \mathcal{O}\left\{\frac{V_o}{[\delta(x_1)]^3}\right\} = \mathcal{O}\left[\frac{V_o Re_1^{3/2}}{x_1^3}\right]. \tag{9.102}$$

Clearly, at $Re_1 >> 1$, the first term on the rhs of (9.101) can be neglected compared to the second term leading to the following simplified version of the equation:

$$\frac{D\omega_3}{Dt} \approx \nu \frac{\partial^2 \omega_3}{\partial x_2^2}. \tag{9.103}$$

Equation (9.103) shows that it is the second gradient of the vorticity in the wall-normal direction amplified by the coefficient of kinematic viscosity that causes the vorticity magnitude to evolve inside the boundary layer region.

9.2 Boundary Layer Separation

We derived the PBLE equation in (9.66) for the flow over a flat plate with the working coordinate system orientated in a manner such that the $x_1(\hat{e}_1)$ axis is parallel to the plate and the $x_2(\hat{e}_2)$ axis is perpendicular to the plate (Fig. 9.1). These equations can also be used to describe the flow field over a curved solid surface (for example, the curved surface of a cylinder, Fig. 9.9) if the following condition is satisfied at all locations on the solid surface:

$$\delta(x_1) << R(x_1), \tag{9.104}$$

where $\delta(x_1)$ is the local boundary layer thickness and $R(x_1)$ is the radius of curvature of the solid surface at x_1 location. If (9.104) holds good, x_1 in the PBLE Eq. (9.66) is interpreted as the coordinate along the curved surface measured from a reference location, and x_2 in that equation is measured from the solid surface along the local

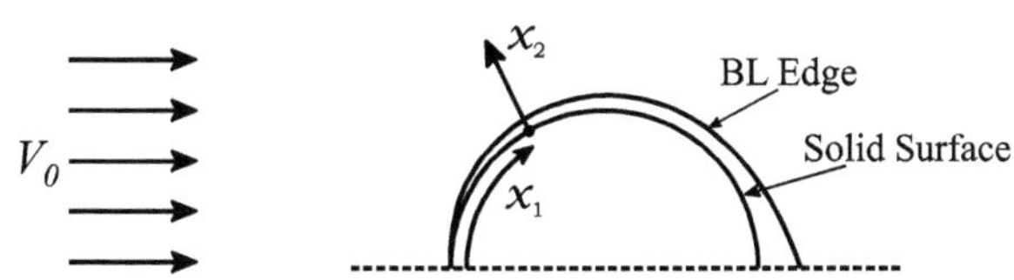

Fig. 9.9 Boundary layer on a curved surface

normal direction. Moreover, for such a curved surface, the definition of the boundary layer thickness needs to be modified

$$\delta(x_1) = x_2 \text{ such that } V_1(x_1, x_2) = 0.99V_t(x_1), \tag{9.105}$$

where $V_t(x_1)$ is the magnitude of the local tangential velocity that would exist on the solid curved surface if $\mu = 0$ (the corresponding potential flow past the solid curved surface). While $V_t(x_1) = V_o$ for the case of a flat plate which is kept parallel to the far-upstream velocity vector, $V_t(x_1) \neq V_o$ for a general curved surface. For example, in the case of solid cylinder, $V_t(x_1) = 2V_o sin\theta'$ (see Eq. 8.75 and Fig. 8.9).

At the edge of the boundary layer, we have

$$\left(V^2\right)_{(x_1,\delta(x_1))} = \left(V_1^2\right)_{(x_1,\delta(x_1))} + \left(V_2^2\right)_{(x_1,\delta(x_1))}. \tag{9.106}$$

However, based on the OM analysis, we are aware that at $Re_1 >> 1$ (Eqs. 9.24 and 9.32)

$$\mathcal{O}\left(\frac{V_2}{V_1}\right) = \mathcal{O}\left(\frac{V_{2C}}{V_{1C}}\right) = \mathcal{O}\left(\frac{L_{2C}}{L_{1C}}\right) = \mathcal{O}\left(\frac{1}{\sqrt{Re_1}}\right) << 1. \tag{9.107}$$

Thus, we approximate (9.106) to

$$\left(V^2\right)_{(x_1,\delta(x_1))} \approx \left(V_1^2\right)_{(x_1,\delta(x_1))}. \tag{9.108}$$

Further, by using the definition of $\delta(x_1)$ (9.105) to express V_1 at the edge of the boundary layer in terms of V_t, Eq. (9.108) is expressed as

$$\left(V^2\right)_{(x_1,\delta(x_1))} \approx (0.99V_t)^2. \tag{9.109}$$

Next, using the generalized Bernoulli Equation (8.40), we find the variation in $p^{edge}(x_1)$ (assuming the body forces to be absent)

$$p^{edge}(x_1) \approx p_o + \frac{\rho(V_o)^2}{2} - \frac{\rho(0.99V_t)^2}{2}, \tag{9.110}$$

where the rhs is a function of x_1. Thus,

$$\frac{dp^{edge}}{dx_1} \approx -\rho(0.99V_t)\frac{dV_t}{dx_1}. \tag{9.111}$$

Since for a curved surface, in general, $\frac{dV_t}{dx_1} \neq 0$, it implies that $\frac{dp^{edge}}{dx_1} \neq 0$, where the partial derivative with respect to x_1 represents the partial derivative with respect to the distance measured along the curved surface. The region on a curved surface where

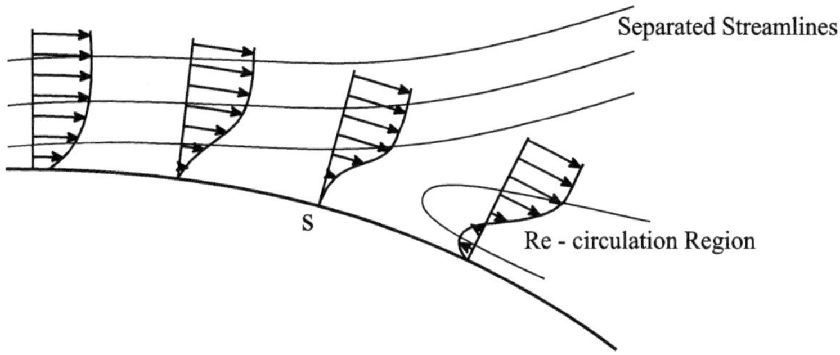

Fig. 9.10 Boundary layer separation

$\frac{dp^{edge}}{dx_1} < 0$ is called the *region of favorable pressure gradient*. The region on a curved surface where $\frac{dp^{edge}}{dx_1} > 0$ is called the *region of adverse pressure gradient*.

While examining the potential flow field past a circular cylinder in Chap. 8 (Fig. 8.9), we observed that on the side of the cylinder surface where a fluid particle experiences increasing blockage ($0 < \theta' < \pi/2$), the pressure gradient is favorable. In contrast, on the side of the solid object where a fluid particle experiences decreasing blockage ($\pi/2 < \theta' < \pi$), the pressure gradient is adverse. Thus, in the region of adverse pressure gradient, fluid particles moving inside the boundary layer experience retardation not only due to the viscous forces but also due to the adverse pressure gradient. With these combined forces of retardation, it is possible that some fluid particles may lose their forward momentum completely and subsequently are forced to reverse their direction of motion. This phenomenon is illustrated in Fig. 9.10. In this figure, we show multiple profiles of V_1 (a component of velocity along the local surface tangent) versus x_2 (the wall-normal distance) at different locations (x_1) on the surface of a cylinder. Due to the additional retardation imposed by the adverse pressure gradient, at location S on the solid surface, $\left(\frac{\partial V_1}{\partial x_2}\right)_{x_1, x_2=0}$ vanishes and at locations further downstream $\left(\frac{\partial V_1}{\partial x_2}\right)_{x_1, x_2=0}$ becomes negative, implying that the fluid particles which are close to the wall are now moving in the opposite direction. This phenomenon is called *boundary layer separation* or *flow separation*, and the location S where $\left(\frac{\partial V_1}{\partial x_2}\right)_{x_1, x_2=0}$ is called the *separation point*. As shown in Fig. 9.10, the streamlines, which have been closely following the curvature of the solid surface up to the point S, tend to move away from the solid surface downstream of the separation point. A new type of flow zone is created between the separated streamlines and the solid surface, where fluid particles tend to translate following closed streamlines. Such a flow zone is called a *recirculation zone*. In this zone $\mathcal{O}[V_2] = \mathcal{O}[V_1]$ and $\mathcal{O}[L1C] = \mathcal{O}[L_{2C}]$. The reticulation zone may have an unsteady or a steady velocity field. The PBLE equation set (9.45) can no longer be used to describe the flow field.

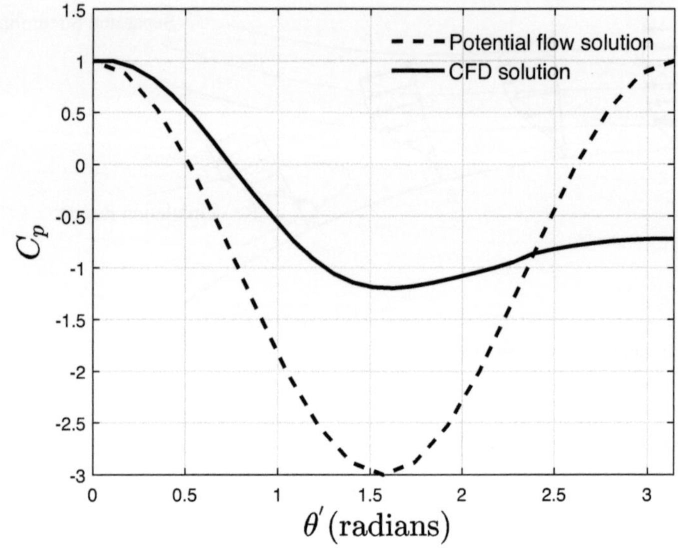

Fig. 9.11 Distribution of C_p on the surface of a cylinder. CFD solution of the full incompressible Navier–Stokes equation set (5.110) at $Re_D = 40$

In Fig. 9.11, we present C_p (8.77) on the surface of a long cylinder employing CFD solution of the full Navier–Stokes Equation (5.110) over the entire flow domain that includes the outer region as well as the boundary layer region (Fig. 9.3). The Reynolds number based on the diameter (d) of the cylinder, Re_d $(= \rho V_o d/\mu)$, for this CFD simulation is 40. In Fig. 9.12, we present some representative streamlines from this numerically simulated flow field. The figure clearly shows the presence of recirculation zones existing on the rear side of the cylinder. In Fig. 9.11, the corresponding variation in C_p as predicted by the potential flow theory (8.78) is also included. We observe that the C_p distribution from the CFD simulation has a distinguishing feature compared to the C_p predicted by the potential flow theory. In the CFD solution, C_p first decreases with increasing θ'. However, at larger θ' values, it tends to become constant. In contrast, in the potential flow solution, the C_p distribution over the rear part of the cylinder surface undergoes a full "recovery" to reach the same value (unity) that it has at the forward stagnation point. The partial "recovery" in C_p shown by the CFD solution always leads to a non-zero positive value when the integral of (6.10) is calculated to determine the pressure drag experienced by the body. In contrast, the C_p distribution predicted using the potential flow theory invariably results in zero pressure drag (this is true not only for a cylinder but for a body of any shape).

As discussed in Sect. 6.2, besides the pressure drag, a solid object interacting with a viscous fluid experiences the viscous drag force (6.11), as well. A body for which the pressure drag is more than the viscous drag is called a *bluff body*, whereas a body for which the pressure drag is less than the viscous drag is called a *streamlined body*.

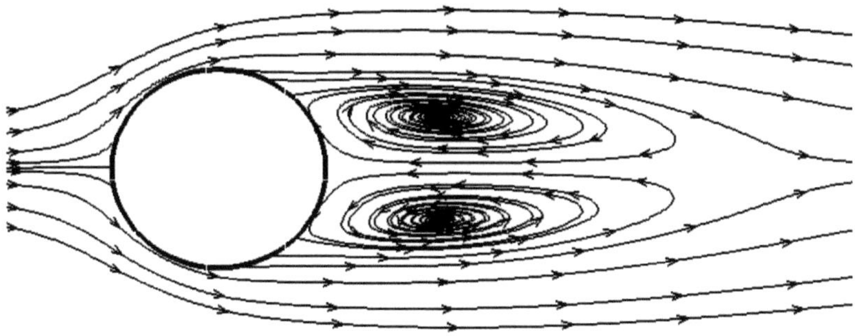

Fig. 9.12 Streamlines in a separated flow past a cylinder. CFD solution of the full incompressible Navier–Stokes equation set (5.110) at $Re_D = 40$

9.3 Internal Flows and Fully Developed Velocity Profiles

Flow fields can be categorized as *external flows* or *internal flows*. An internal flow is one in which boundary layer effects spread from multiple directions in the flow domain. The flow between two parallel plates (as shown in Fig. 9.13) is an example of internal flows. Here, the boundary layer effects spread in the flow field, emanating both from the top and the bottom plates. In contrast, in external flows, the boundary layer effects spread in the flow field, emanating from only one side. The flow past a flat plate (Fig. 9.1) and the flow past a cylinder (Fig. 8.9) are examples of external flows.

In this section, we focus on the flow between two horizontal flat plates. The separation between the plates is H. This setup is exposed to a uniform stream, as shown in Fig. 9.13. We assume the flow between the plates to be laminar and steady, and the body force is negligible. Further, the plates are assumed to be very wide in the $\hat{e}_3$ direction such that the velocity field between the plates is 2C and 2D. As the uniform stream starts interacting with the plate surfaces, individual boundary layers develop over the two plates. At stream-wise stations (x_1) where the individual boundary layer thickness is still less than $H/2$ (half the separation between the plates), there

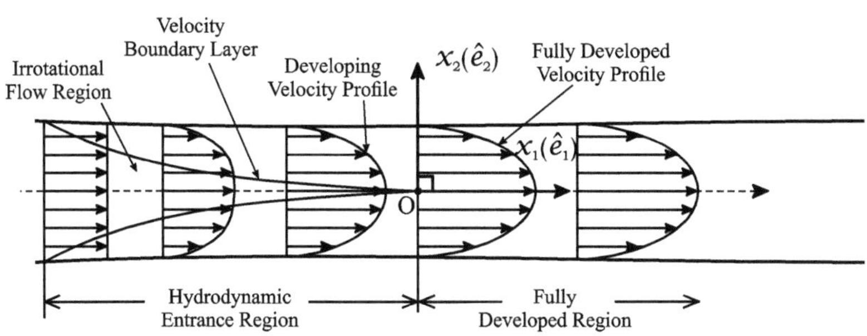

Fig. 9.13 Internal flow between two parallel plates

still exists a vorticity-free flow region (as shown in Fig. 9.13). Some representative velocity profiles of the stream-wise velocity versus the wall-normal distance are also sketched in Fig. 9.13. As one moves further downstream, this vorticity-free zone continuously shrinks and eventually disappears beyond the point where the individual boundary layer thickness becomes equal to H. The distance of this point measured from the left edge of one of the plates is called the *hydrodynamic entrance length* (L_e in Fig. 9.13). It is experimentally observed that at all locations downstream of this point, the profile of stream-wise velocity remains unchanged. Such a non-changing velocity profile is called a *fully developed velocity profile*. In a ground-fixed Cartesian coordinate system $Ox_1(\hat{e}_1)x_2(\hat{e}_2)x_3(\hat{e}_3)$ (as shown in Fig. 9.13), it means

$$\frac{\partial V_1}{\partial x_1} = 0. \tag{9.112}$$

Our aim in this section is to leverage (9.112) and further simplify the steady 2C, 2D incompressible Navier–Stokes equation set (Eqs. 9.3–9.5), so that we can arrive at the exact solution of the flow field. This derived solution may offer additional opportunities to gain further insights into internal flows and the role of no-slip conditions in shaping flow behavior. In the region of fully developed V_1 profile ($x_1 > 0$) the continuity equation (9.3) simplifies to

$$\frac{\partial V_2}{\partial x_2} = 0, \tag{9.113}$$

which in turn implies

$$V_2(x_1, x_2) = C, \tag{9.114}$$

where either C is a constant or it is a function of x_1. However, the no-penetration condition implies $V_2(x_1, x_2 = -H/2) = 0$ at all x_1 stations. Thus, C must be zero, and consequently, V_2 must be zero everywhere in the flow field.

$$V_2(x_1, x_2) = 0 \tag{9.115}$$

rendering the 1C velocity field

$$\underline{V} = V_1(x_2)\hat{e}_1. \tag{9.116}$$

The conclusion in (9.116) can be used to verify that every fluid particle in this steady velocity field has zero acceleration.

$$\frac{D\underline{V}}{Dt} = \frac{\partial \underline{V}}{\partial t} + \left[V_1\frac{\partial V_1}{\partial x_1} + V_2\frac{\partial V_1}{\partial x_2}\right]\hat{e}_1 + \left[V_1\frac{\partial V_2}{\partial x_1} + V_2\frac{\partial V_2}{\partial x_2}\right]\hat{e}_2 = 0. \tag{9.117}$$

With V_1 being the only non-zero component of velocity, every fluid particle follows a path-line which is straight and oriented exactly parallel to the plates. These flow features are different from what we observed earlier in the Blassius solution. Inside the boundary layer of a single flat plat, the velocity field had non-zero V_2, albeit small. The streamlines do bend there ($V_1 \neq 0$ and $V_2 \neq 0$), and fluid particles do accelerate, as well.

In the fully developed region of Fig. 9.13, since fluid particles do not accelerate, the V_1 Eq. (9.4) simplifies to a balance between the pressure gradient force and the viscous forces.

$$0 = -\frac{1}{\rho}\frac{\partial p}{\partial x_1} + \nu \frac{\partial^2 V_1}{\partial x_2 \partial x_2}. \tag{9.118}$$

Similarly, the V_2 Eq. (9.5) reduces merely to the conclusion pressure is not a function of x_2

$$0 = -\frac{1}{\rho}\frac{\partial p}{\partial x_2}. \tag{9.119}$$

Equation (9.119) in conjunction with the already simplified version of the V_3 Eq. (9.6) leads to the conclusion that pressure must be a function of x_1 alone or a constant.

$$p = p(x_1) \text{ or } p = p_o(\text{a constant}). \tag{9.120}$$

Recognizing that V_1 depends on x_2 alone and that p must depend on x_1 alone (if it is still a variable), the partial derivatives in (9.118) can be substituted by the total derivative operators.

$$\frac{1}{\rho}\frac{dp}{dx_1} = \nu \frac{d^2 V_1}{dx_2 dx_2}. \tag{9.121}$$

Based on (9.121), we rule out the possibility that p can be a constant. In that case, the rhs of (9.121) would be zero, leading to a linear velocity profile which would contradict the no-slip boundary condition that must exist at $x_1 = \pm H/2$. This leaves us with the conclusion that p must be a function of x_1 alone. Further, we know that the rhs of (9.121) is not a function of x_1 (fully developed velocity profile). Thus, we conclude that both the lhs and rhs of (9.121) must be a constant. That constant is the imposed pressure gradient that drives the flow field. Using the symbol λ to represent the constant $\frac{dp}{dx_1}$ (9.121) is expressed as

$$\mu \frac{d^2 V_1}{dx_2 dx_2} = \lambda. \tag{9.122}$$

This is a second-order ordinary differential equation that can be readily integrated to arrive at the exact solution

$$V_1 = \frac{\lambda}{2\mu} x_2^2 + A x_2 + B, \tag{9.123}$$

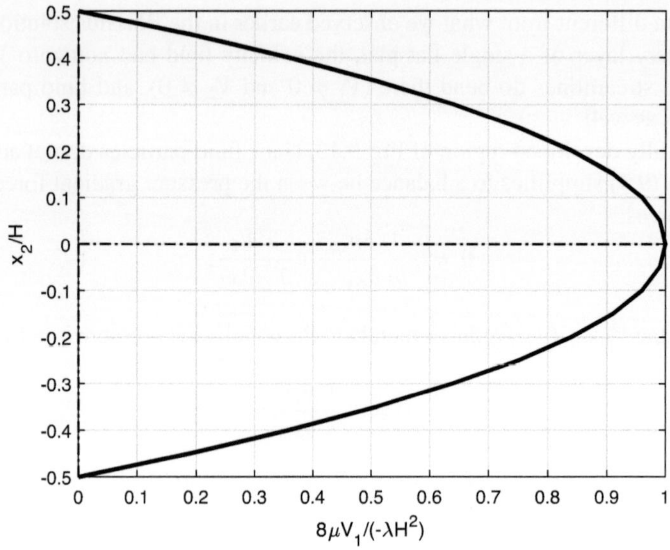

Fig. 9.14 Fully developed parabolic velocity profile in the internal flow between two parallel plates

where A and B are constants of integration. These constants can be determined by imposing the two boundary conditions on V_1

$$V_1(x_2 = -H/2) = 0 \text{ and } V_1(x_2 = H/2) = 0, \tag{9.124}$$

leading to the solution

$$V_1 = \frac{-\lambda H^2}{8\mu}\left[1 - 4\left(\frac{x_2}{H}\right)^2\right]. \tag{9.125}$$

This is a parabolic profile, which is shown in Fig. 9.14, where we have plotted the non-dimensional velocity $\frac{8\mu V_1(x_2)}{-\lambda H^2}$ versus x_2. The profile is symmetric about the x_1 $(\hat{e}_1)$ axis. This flow field is driven by a negative pressure gradient ($\lambda < 0$, high pressure on the left side and low pressure on the right side).

Using (9.125), we find the vorticity in the flow domain.

$$\omega_3 = \frac{\partial V_2}{\partial x_1} - \frac{\partial V_1}{\partial x_2} = \frac{-\lambda x_2}{\mu}. \tag{9.126}$$

The vorticity is non-zero everywhere except on the x_1 $(\hat{e}_1)$ axis. At locations with $x_2 > 0$, the vorticity is positive (when $\lambda < 0$), implying that small local fluid elements therein are rotating in the anticlockwise sense. On the other hand, vorticity is negative (when $\lambda < 0$) at locations with $x_2 < 0$, implying that the local fluid elements therein are rotating in the clockwise manner. The vorticity is most negative at the lower plate ($x_2 = -H/2$) and is most positive at the upper plate ($x_2 = H/2$). Further, the expression in (9.126) also shows that the vorticity varies with x_2 and has no dependence on x_1. We realize that following a small fluid element, not only the acceleration of the center of the mass of the fluid element remains zero in this flow field (9.117), but its vorticity also remains constant. This observation is indeed in line with the governing equation of vorticity in a 2C, 2D velocity field

$$\frac{D\omega_3}{Dt} = \nu \left(\frac{\partial^2 \omega_3}{\partial x_1^2} + \frac{\partial^2 \omega_3}{\partial x_2^2} \right), \tag{9.127}$$

wherein the first term of the rhs vanishes because ω_3 in (9.126) is not dependent on x_1 at all, and the second term of the rhs vanishes because ω_3 in (9.126) is merely a linear function of x_2. This behavior is in contrast to the Blassius solution of flow past a single flat plate, wherein the second term on the rhs of (9.127) is still active inside the boundary layer and is instrumental in introducing vorticity to all constituent fluid particles as they enter the boundary layer region from an outside potential zone (Fig. 9.3).

Chapter 10
Governing Equations of Fluid Motion in Non-inertial Reference Frames

So far in this book, we have considered flows past solid bodies which are fixed to an inertial reference frame (the ground frame). Such solid bodies are mere extensions of the ground frame. The velocity vector appearing in the governing equations of motion (the Navier–Stokes equations) that we derived earlier in Chap. 5 (5.110) is the velocity of a fluid particle as observed by an observer who is fixed to the ground frame. Examples of such flow fields are the flow of river water past a rock, which is fixed to the ground frame, or the flow of wind past buildings. Many engineering problems of interest, however, involve objects moving with respect to the ground frame in an otherwise stationary (with respect to the ground frame) fluid medium. For example, a submarine moving in the otherwise stagnant water of a lake. Often, for such problems, the object itself is taken as the reference frame. However, such a frame, in general, may have combined translational and rotational motion with respect to the ground reference frame. Thus, such an object-fixed reference frame is *non-inertial*, and the governing equation of fluid motion observed with respect to such a frame would be different from the Navier–Stokes equation set derived earlier (5.110). In the next section, we derive the appropriately modified governing equation of fluid motion with respect to a non-inertial reference frame.

10.1 A Non-inertial Reference Frame

Figure 10.1 shows a submarine (the solid body S) which is moving in otherwise stagnant water with respect to the inertial ground frame. The symbols S and G denote the submarine-fixed reference frame and the inertial ground frame, respectively. The symbols C and O represent two points which are fixed to the submarine body and the ground frame, respectively. P represents an arbitrary location in the fluid medium. The instantaneous position vectors of location P measured from point O and point C are $\underline{X}$ and $\underline{X}^{'}$, respectively. The Eulerian variables $V(\underline{X}, t)$ and $V^{'}(\underline{X}^{'}, t)$ denote the

© The Author(s), under exclusive license to Springer Nature Switzerland AG 2026

S. S. Sinha, *Fundamentals of Fluid Mechanics*,
https://doi.org/10.1007/978-3-032-04210-1_10

Fig. 10.1 A translating and rotating object (the solid body S) being used as a reference frame. P is an arbitrary location in the flow domain

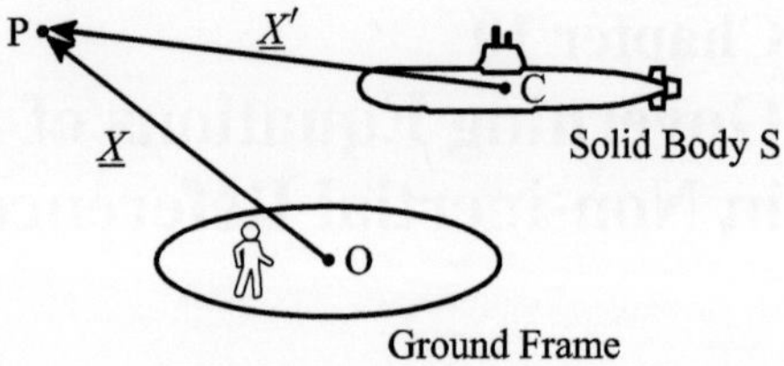

velocity fields of the fluid medium with respect to the frames G and S, respectively. In general, $V(\underline{X}, t) \neq V^{'}(\underline{X}^{'}, t)$, even though both quantities describe the velocity vector of the same fluid particle which is currently located at location P. Equation set (5.110) is the governing equation set of $V(\underline{X}, t)$. We wish to derive the governing equation set of $V^{'}(\underline{X}^{'}, t)$.

The motion of the submarine (which is assumed to be a rigid body) with respect to the inertial ground frame is described completely by the following kinematic quantities, which we assume to be known at the current time instant.

1. $\underline{V}_{C|G}$: Velocity of point C with respect to the ground frame,
2. $\underline{a}_{C|G}$: acceleration of point C with respect to the ground frame,
3. $\underline{\omega}_{S|G}$: Angular velocity of the submarine with respect to the ground frame,
4. $\underline{\dot{\omega}}_{S|G}$: Angular acceleration of the submarine with respect to the ground frame.

The vectors $\underline{V}(\underline{X}, t)$ and $\underline{V}^{'}(\underline{X}^{'}, t)$ are related to each other through a so-called the *velocity-transfer relationship* (see Dumir et al. [1] for details).

$$\underline{V}(\underline{X}, t) = \underline{V}^{'}(\underline{X}^{'}, t) + \underline{\omega}_{S|G} \times \underline{X}^{'} + \underline{V}_{C|G}. \tag{10.1}$$

Let the symbols $\underline{a}(\underline{X}, t)$ and $\underline{a}^{'}(\underline{X}^{'}, t)$ denote the acceleration vectors of the same fluid particle (currently located at P), but with respect to different reference frames. In general $\underline{a}(\underline{X}, t) \neq \underline{a}^{'}(\underline{X}^{'}, t)$. These vectors can be expressed in terms of appropriate material derivatives of appropriate velocity vectors

$$\underline{a}(\underline{X}, t) = \left\{\frac{\partial}{\partial t} + \left[\underline{\nabla} \cdot \underline{V}(\underline{X}, t)\right]\right\} \underline{V}(\underline{X}, t), \tag{10.2}$$

and

$$\underline{a}^{'}(\underline{X}^{'}, t) = \left\{\frac{\partial}{\partial t} + \left[\underline{\nabla} \cdot \underline{V}^{'}(\underline{X}^{'}, t)\right]\right\} \underline{V}^{'}(\underline{X}^{'}, t). \tag{10.3}$$

These two vectors, which are the acceleration vectors of the same particle, are related by the so-called *acceleration transfer relationship* (see Dumir et al. [1] for details).

$$\begin{aligned}\underline{a}(\underline{X}, t) = {} & \underline{a}^{'}(\underline{X}^{'}, t) + \underline{\dot{\omega}}_{S|G} \times \underline{X}^{'} + \underline{\omega}_{S|G} \times \left(\underline{\omega}_{S|G} \times \underline{X}^{'}\right) \\ & + 2\underline{\omega}_{S|G} \times \underline{V}^{'}(\underline{X}^{'}, t) + \underline{a}_{C|G}.\end{aligned} \tag{10.4}$$

The first term on the rhs is the acceleration of the fluid particle as observed with respect to the submarine frame. This is the same as (10.3). The second term on the rhs is the cross product between the angular acceleration of the submarine with respect to the ground frame and the instantaneous position vector of location P measured from point C. The third term on rhs is a double cross product involving the angular velocity of the submarine and the instantaneous position vector $\underline{X}^{'}$ of the location measured from point C. The term is commonly referred to as the *centripetal acceleration*. The fourth term on the rhs of (10.4) is twice the cross product of the angular acceleration of the submarine and the instantaneous velocity vector of the particle with respect to the submarine frame. This term is commonly referred to as the *Coriolis acceleration*. Of these four terms, the first term can be substituted in terms of $\underline{V}^{'}\left(\underline{X}^{'}, t\right)$ using (10.3) leading to the following form of (10.4)

$$\begin{aligned}\underline{a}(\underline{X}, t) = &\left[\frac{\partial}{\partial t} + \underline{\nabla} \cdot \underline{V}^{'}(\underline{X}^{'}, t)\right] \underline{V}^{'}(\underline{X}^{'}, t) + \underline{\dot{\omega}}_{S|G} \times \underline{X}^{'} \\ &+ \underline{\omega}_{S|G} \times \left(\underline{\omega}_{S|G} \times \underline{X}^{'}\right) + 2\underline{\omega}_{S|G} \times \underline{V}^{'}(\underline{X}^{'}, t) \\ &+ \underline{a}_{C|G}.\end{aligned} \tag{10.5}$$

This must equal the net force per unit mass experienced by the fluid particle currently located at P.

From our earlier derivation of the Navier–Stokes equation set (5.110), we already know that the net force per unit mass ($\underline{f}^{net}$) equals

$$\underline{f}^{net}(\underline{X}, t) = \underline{g} - \frac{\underline{\nabla} p}{\rho} + \nu \nabla^2 \underline{V}(\underline{X}, t). \tag{10.6}$$

Using the rhs of (10.6) to substitute for $\underline{a}(\underline{X}, t)$ in (10.5) leads to the following equation

$$\begin{aligned}\underline{g} - \frac{\underline{\nabla} p\left(\underline{X}, t\right)}{\rho} + \nu \nabla^2 \underline{V}\left(\underline{X}, t\right) = &\left[\frac{\partial}{\partial t} + \underline{\nabla} \cdot \underline{V}^{'}(\underline{X}^{'}, t)\right] \underline{V}^{'}(\underline{X}^{'}, t) \\ &+ \underline{\dot{\omega}}_{S|G} \times \underline{X}^{'} + \underline{\omega}_{S|G} \times \left(\underline{\omega}_{S|G} \times \underline{X}^{'}\right) \\ &+ 2\underline{\omega}_{S|G} \times \underline{V}^{'}(\underline{X}^{'}, t) + \underline{a}_{C|G}.\end{aligned} \tag{10.7}$$

On the lhs of (10.7), the independent spatial variable is still $\underline{X}$. However, since $\underline{X}^{'}$ and $\underline{X}$ both point to the same physical location (P),

$$\underline{\nabla} p\left(\underline{X}, t\right) = \underline{\nabla} p\left(\underline{X}^{'}, t\right). \tag{10.8}$$

Further, using (10.1) to substitute the vector $\underline{V}(\underline{X}, t)$ by vector $\underline{V}^{'}(\underline{X}^{'}, t)$ in the viscous force per unit mass term on the rhs of (10.6), we have

$$\begin{aligned}\nabla^2 \underline{V}(\underline{X}, t) &= \nabla^2 \left[\underline{V}^{'}(\underline{X}^{'}, t) + \underline{\omega}_{S|G} \times \underline{X}^{'} + \underline{V}_{C|G}\right], \\ &= \nabla^2 \left[\underline{V}^{'}(\underline{X}^{'}, t)\right] + \nabla^2 \left[\underline{\omega}_{S|G} \times \underline{X}^{'}\right] + \nabla^2 \left[\underline{V}_{C|G}\right].\end{aligned} \tag{10.9}$$

In (10.9), the second term on the rhs must vanish because the ∇^2 operator is acting on a linear function of $\underline{X}^{'}$. The last term on the rhs must vanish because the ∇^2 operator acts on a spatially constant vector ($\underline{V}_{C|G}$). Thus,

$$\nabla^2 \underline{V}(\underline{X}, t) = \nabla^2 \underline{V}^{'}(\underline{X}^{'}, t). \tag{10.10}$$

Substituting (10.8) and (10.10) on the lhs of (10.7) and re-arranging the resulting equation, we arrive at the following equation wherein all independent and dependent variables are now completely described with reference to the submarine frame.

$$\begin{aligned}\left[\frac{\partial}{\partial t} + \underline{\nabla} \cdot \underline{V}^{'}(\underline{X}^{'}, t)\right] \underline{V}^{'}(\underline{X}^{'}, t) &= \underline{g} - \frac{\underline{\nabla} p\left(\underline{X}^{'}, t\right)}{\rho} + \nu \nabla^2 \underline{V}^{'}(\underline{X}^{'}, t) \\ &\quad - \dot{\underline{\omega}}_{S|G} \times \underline{X}^{'} - \underline{\omega}_{S|G} \times \left(\underline{\omega}_{S|G} \times \underline{X}^{'}\right) \\ &\quad - 2\underline{\omega}_{S|G} \times \underline{V}^{'}(\underline{X}^{'}, t) - \underline{a}_{C|G}.\end{aligned} \tag{10.11}$$

This is the appropriate form of Euler's first axiom when a non-inertial frame is used to describe the motion of a fluid medium.

Further, using (10.1), the continuity equation (5.110) can also be recast with $\underline{V}^{'}(\underline{X}^{'}, t)$ as the dependent variable. We first express the divergence of $\underline{V}(\underline{X}, t)$.

$$\begin{aligned}\underline{\nabla} \cdot \left[\underline{V}(\underline{X}, t)\right] &= \underline{\nabla} \cdot \left[\underline{V}^{'}(\underline{X}^{'}, t) + \underline{\omega}_{S|G} \times \underline{X}^{'} + \underline{V}_{C|G}\right], \\ &= \underline{\nabla} \cdot \left[\underline{V}^{'}(\underline{X}^{'}, t)\right] + \underline{\nabla} \cdot \left[\underline{\omega}_{S|G} \times \underline{X}^{'}\right] \\ &\quad + \underline{\nabla} \cdot \left[\underline{V}_{C|G}\right].\end{aligned} \tag{10.12}$$

The third term on the rhs of (10.12) vanishes because the divergence operator acts on a constant vector ($\underline{V}_{C|G}$). To understand the consequence of the second term on the rhs, we express the quantity using a working Cartesian coordinate system $Cx_1^{'}(\hat{e}_1^{'})x_2^{'}(\hat{e}_2^{'})x_3^{'}(\hat{e}_3^{'})$, which is fixed to the submarine and has its origin at C. The three mutually perpendicular unit vectors of this coordinate system are $\hat{e}_1^{'}$, $\hat{e}_2^{'}$, and $\hat{e}_3^{'}$, respectively. The symbols $x_1^{'}$, $x_2^{'}$ and $x_3^{'}$ denote the coordinates of the arbitrary location P, such that,

$$\underline{X}^{'} = x_q^{'} \hat{e}_q^{'}, \nabla = \hat{e}_m^{'} \frac{\partial}{\partial x_m^{'}}, \omega_{S|G} = \omega_p \hat{e}_p^{'}. \tag{10.13}$$

Thus,

$$\underline{\nabla} \cdot \left[\underline{\omega}_{S|G} \times \underline{X}^{'}\right] = \hat{e}^{'}_{m} \frac{\partial}{\partial x^{'}_{m}} \cdot \left[\varepsilon_{pqr}\omega_{p}x^{'}_{q}\hat{e}^{'}_{r}\right],$$
$$= \hat{e}^{'}_{m} \cdot \hat{e}^{'}_{r}\varepsilon_{pqr}\omega_{p}\frac{\partial x^{'}_{q}}{\partial x^{'}_{m}} = \delta_{mr}\varepsilon_{pqr}\omega_{p}\delta_{qm},$$
$$= \delta_{mr}\varepsilon_{pmr}\omega_{p} = \varepsilon_{pmm}\omega_{p}, \tag{10.14}$$

where $\partial x^{'}_{q}/\partial x^{'}_{m}$ has been aptly replaced by δ_{qm}. We have

$$\varepsilon_{pmm}\omega_{p} = \varepsilon_{p11}\omega_{p} + \varepsilon_{p22}\omega_{p} + \varepsilon_{p33}\omega_{p}. \tag{10.15}$$

However, by the definition of the permutation symbol (1.46),

$$\varepsilon_{ijk} = 0 \text{ if } j = k. \tag{10.16}$$

Thus, $\varepsilon_{pmm}\omega_{p} = 0$. This, in turn, implies

$$\underline{\nabla} \cdot \left[\underline{\omega}_{S|G} \times \underline{X}^{'}\right] = 0. \tag{10.17}$$

Thus Eq. (10.12) becomes

$$\underline{\nabla} \cdot \left[\underline{V}(\underline{X}, t)\right] = \underline{\nabla} \cdot \left[\underline{V}^{'}(\underline{X}^{'}, t)\right]. \tag{10.18}$$

Using (10.18) in the continuity equation (5.110) leads to the representation of the continuity equation in terms of $\underline{V}(\underline{X}^{'}, t)$.

$$\underline{\nabla} \cdot \left[\underline{V}^{'}(\underline{X}^{'}, t)\right] = 0. \tag{10.19}$$

In summary, (10.11) and (10.19) are the governing equations of motion of a constant-density fluid with respect to the non-inertial submarine-fixed reference frame

$$\underline{\nabla} \cdot \left[\underline{V}^{'}(\underline{X}^{'}, t)\right] = 0,$$
$$\left[\frac{\partial}{\partial t} + \underline{\nabla} \cdot \underline{V}^{'}(\underline{X}^{'}, t)\right]\underline{V}^{'}(\underline{X}^{'}, t) = \underline{g} - \frac{\underline{\nabla} p\left(\underline{X}^{'}, t\right)}{\rho} + \nu\nabla^{2}\underline{V}^{'}(\underline{X}^{'}, t)$$
$$- \dot{\underline{\omega}}_{S|G} \times \underline{X}^{'} - \underline{\omega}_{S|G} \times \left(\underline{\omega}_{S|G} \times \underline{X}^{'}\right)$$
$$- 2\underline{\omega}_{S|G} \times \underline{V}^{'}(\underline{X}^{'}, t) - \underline{a}_{C|G}. \tag{10.20}$$

Since our reference frame is fixed to the submarine body, the no-slip and the no-penetration boundary conditions on the surface of the submarine are to be expressed accordingly. The no-slip condition is

$$\hat{e}_n \cdot \underline{V}'(\underline{X}', t) = 0, \tag{10.21}$$

and the no-penetration condition is

$$\underline{V}'(\underline{X}', t) - \hat{e}_n \cdot \underline{V}'(\underline{X}', t) = 0, \tag{10.22}$$

where $\hat{e}_n$ denotes the local outward normal unit vector on the submarine surface.

Further, using $Cx_1'(\hat{e}_1')x_2'(\hat{e}_2')x_3'(\hat{e}_3')$ as our working coordinate system and then using the index notation to express various tensors, we can demonstrate that

$$\frac{[\underline{\nabla}\,\underline{V}(\underline{X}, t)] + [\underline{\nabla}\,\underline{V}(\underline{X}, t)]^T}{2} = \frac{\left[\underline{\nabla}\,\underline{V}'(\underline{X}', t)\right] + \left[\underline{\nabla}\,\underline{V}'(\underline{X}', t)\right]^T}{2}, \tag{10.23}$$

and

$$\frac{[\underline{\nabla}\,\underline{V}(\underline{X}, t)] - [\underline{\nabla}\,\underline{V}(\underline{X}, t)]^T}{2} = \frac{\left[\underline{\nabla}\,\underline{V}'(\underline{X}', t)\right] - \left[\underline{\nabla}\,\underline{V}'(\underline{X}', t)\right]^T}{2} + \underline{\nabla}\left(\underline{\omega}_{S|G} \times \underline{X}'\right). \tag{10.24}$$

The lhs of (10.23) is the local strain-rate tensor with respect to frame G, whereas the rhs is the local strain-rate tensor with respect to frame F (the symmetric parts of the relevant velocity gradient tensor). Equation (10.23) implies that the strain-rate tensor is a frame-independent quantity. Equation (10.23) also implies that the viscous stress tensor at any location in a Newtonian fluid can be directly obtained using the gradient of the measured/computed velocity field $\underline{V}'(\underline{X}', t)$ itself,

$$\underline{\tau} = 2\mu\frac{[\underline{\nabla}\,\underline{V}(\underline{X}, t)] + [\underline{\nabla}\,\underline{V}(\underline{X}, t)]^T}{2} = 2\mu\frac{\left[\underline{\nabla}\,\underline{V}'(\underline{X}', t)\right] + \left[\underline{\nabla}\,\underline{V}'(\underline{X}', t)\right]^T}{2}. \tag{10.25}$$

In contrast, (10.24) demonstrates the rotation-rate tensors (the antisymmetric part of the velocity gradient tensor) computed using $\underline{V}(\underline{X}, t)$ and $\underline{V}'(\underline{X}', t)$ are different. The difference depends on $\underline{\omega}_{S|G}$. Thus, the rotation-rate tensor is a frame-dependent quantity.

10.2 A Frame Translating with Constant Velocity

Many engineering applications involve flow fields wherein an object of interest (a submarine or an aircraft) is moving with a constant, known velocity with respect to the ground frame in a fluid medium which is otherwise stationary with respect to the ground frame. For such flow fields,

1. The velocity of every material solid point on the submarine with respect to the ground frame is identical, and that equals $\underline{V_o}$. Thus, $\underline{V}_{C|G} = \underline{V_o}$, where C is a point fixed on the submarine.
2. Acceleration of every point on the submarine with respect to the ground frame is 0. Thus, $\underline{a}_{C|G} = 0$.
3. Angular velocity of the submarine with respect to the ground frame is zero ($\underline{\omega}_{S|G} = 0$).
4. Angular acceleration of the submarine with respect to the ground frame is 0 ($\underline{\dot{\omega}}_{S|G} = 0$).

Thus, the governing equation set of fluid motion, as observed with respect to the submarine reference frame, is obtained by appropriately simplifying (10.20).

$$\nabla \cdot \left[\underline{V}^{'}(\underline{X}^{'}, t)\right] = 0,$$
$$\left[\frac{\partial}{\partial t} + \underline{\nabla} \cdot \underline{V}^{'}(\underline{X}^{'}, t)\right] \underline{V}^{'}(\underline{X}^{'}, t) = \underline{g} - \frac{\underline{\nabla} p\left(\underline{X}^{'}, t\right)}{\rho} + \nu \nabla^2 \underline{V}^{'}(\underline{X}^{'}, t). \tag{10.26}$$

Equation set (10.26) appears similar to the Navier–Stokes equation set derived with respect to the ground frame (5.110): there are no additional acceleration terms therein. This is not unexpected. After all, a frame translating with a constant velocity with respect to an inertial frame is itself inertial. However, the important difference in (10.26) compared to (5.110) is that the relevant velocity vector is $\underline{V}^{'}(\underline{X}^{'}, t)$ and the relevant spatial variable is $\underline{X}^{'}$. The associated boundary conditions on the surface of the submarine are as described in (10.21). Further, since the fluid is stagnant with respect to the ground frame in the far-upstream region, the corresponding boundary condition on $\underline{V}^{'}(\underline{X}^{'}, t)$ in the far-upstream region must be

$$\underline{V}^{'}(\underline{X}^{'}, t) = -\underline{V_o}. \tag{10.27}$$

The rhs of (10.27) is the negative of the constant velocity vector with which the object is translating with respect to the ground frame.

References

1. S. Sengupta P. C. Dumir and S. V. Veeravalli. Engineering Mechanics. University Press (India) Private Limited, Hyderabad, India, 2020, pp. 24–26.
2. J. N. Reddy. An Introduction to Continuum Mechanics. Cambridge English, 2015.
3. E. L. Houghton and N.B. Carruthers. Aerodyanmics for Engineering Students. Hodder Arnold, 3rd Edition, 1982.

© The Author(s), under exclusive license to Springer Nature Switzerland AG 2026
S. S. Sinha, *Fundamentals of Fluid Mechanics*,
https://doi.org/10.1007/978-3-032-04210-1

Index

A
Advection operator, 14
Adverse pressure gradient, 126, 157
Antisymmetric tensor, 5
Averaged angular velocity vector, 42

B
Bernoulli's equation, 104
Blassius solution, 144, 148
Bluff body, 158
Body forces, 61
Boundary layer separation, 155
Boundary layer thickness, 128
Buoyancy force, 99

C
Center of curvature, 105
Characteristic length scale, 131
Characteristic value of an Eulerian variable, 130
Coefficient of dynamic viscosity, 81
Continuity equation, 59
Continuum description, 27
Control volume, 53
Coriolis acceleration, 167
Cross product, 12
Curl of a vector, 14
Cylindrical-polar coordinate system, 19

D
Differential form of the Reynolds transport theorem, 55
Dilatation-rate, 47
Divergence of a tensor, 13
Dot product, 9
Double dot product, 11
Doublet, 117
Drag force, 90
Dummy index, 7
Dyad, 4
Dynamic pressure, 122

E
Einstein's summation rule, 6
Eulerian description, 30
Euler's equation set, 104
Euler's first axiom, 26, 60, 168
Euler's second axiom, 26
External flows, 159

F
Favorable pressure gradient, 125, 157
First law of thermodynamics, 72
Flat plate boundary layer, 127
Fluid element, 38
Fluid particle, 27

© The Author(s), under exclusive license to Springer Nature Switzerland AG 2026

S. S. Sinha, *Fundamentals of Fluid Mechanics*,
https://doi.org/10.1007/978-3-032-04210-1

Flux through a surface, 51
Fourier's law of heat conduction, 82
Free index, 7
Free vortex, 116
Froude number, 136
Fully developed velocity profile, 160

G
Generalized Bernoulli's equation, 113
Gradient of a tensor, 13

H
Hydrodynamic entrance length, 160
Hydrostatics, 93

I
Ideal fluid, 103
Incompressible Navier–Stokes equation set, 84
Index notation, 8
Internal energy, 53
Internal flows, 159

K
Kronecker delta, 11

L
Lagrangian description, 29
Laminar flows, 127
Laplacian of a tensor, 14
Lift force, 90

M
Material derivative, 33

N
Navier–Stokes equation set, 83
Newtonian fluids, 79
Newton's third law, 90
Non-inertial reference frame, 165
Non-Newtonian fluids, 82
No-penetration, 88
Normal strain-rate components, 49

O
Order-of-magnitude, 130

P
Panel methods, 126
Path coordinate system, 104
Pathline, 36
Permutation symbol, 12
Potential flows, 111
Potential function of a free vortex, 116
Potential function of a source/sink, 115
Prandtl's boundary layer equation set, 139
Pressure-dilatation, 83
Pressure drag, 90
Pressure–stress tensor, 79
Principal normal unit vector, 105

R
Recirculation zone, 157
Reynolds number, 134
Reynolds transport theorem, 53
Rotation-rate tensor, 48

S
Self-similarity, 140
Separation point, 157
Shear strain-rate components, 49
Source/sink, 115
Spatial derivatives of tensors, 13
Stagnation points, 124
Steady velocity field, 35
Strain-rate tensor, 48
Streakline, 36
Stream function, 140
Streamline, 36
Streamlined body, 158
Stress tensor, 61
Superposition of potential functions, 120
Surface forces, 61
Symmetric tensor, 5

T
Tangent unit vector, 105
Tensor, 1
Tensor identities, 15
Transformation rule of first-order tensors, 2
Transformation rule of second-order tensor, 5
Transitional flows, 127
Transport equation, 59
Transpose of a second-order tensor, 4
Turbulent flows, 127

U
Uniform velocity field, 35

V
Viscous drag, 90
Viscous fluids, 127
Viscous heating, 83
Viscous stress tensor, 79
Volumetric heating, 74
Vorticity, 163
Vorticity equation, 109
Vorticity vector, 48

MIX
Papier aus verantwortungsvollen Quellen
Paper from responsible sources
FSC® C105338

If you have any concerns about our products,
you can contact us on
ProductSafety@springernature.com

In case Publisher is established outside the EU,
the EU authorized representative is:
Springer Nature Customer Service Center GmbH
Europaplatz 3, 69115 Heidelberg, Germany

Printed by Libri Plureos GmbH
in Hamburg, Germany